PLANNERS AND LOCAL POLITICS
Impossible Dreams

Volume 7, Sage Library of Social Research

SAGE LIBRARY OF SOCIAL RESEARCH

Also in this series:

1. **DAVID CAPLOVITZ**
   The Merchants of Harlem: A Study of Small Business in a Black Community
2. **JAMES N. ROSENAU**
   International Studies and the Social Sciences: Problems, Priorities and Prospects in the United States
3. **DOUGLAS E. ASHFORD**
   Ideology and Participation
4. **PATRICK J. McGOWAN and HOWARD B. SHAPIRO**
   The Comparative Study of Foreign Policy: A Survey of Scientific Findings
5. **GEORGE A. MALE**
   The Struggle for Power: Who Controls the Schools in England and the United States
6. **RAYMOND TANTER**
   Modelling and Managing International Conflicts: The Berlin Crises
7. **ANTHONY JAMES CATANESE**
   Planners and Local Politics: Impossible Dreams
8. **JAMES RUSSELL PRESCOTT**
   Economic Aspects of Public Housing
9. **F. PARKINSON**
   Latin America, the Cold War, and the World Powers, 1945-1973: A Study in Diplomatic History
10. **ROBERT G. SMITH**
    Ad Hoc Governments: Special Purpose Transportation Authorities in Britain and the United States
11. **RONALD GALLIMORE, JOAN WHITEHORN BOGGS, and CATHIE JORDAN**
    Culture, Behavior and Education: A Study of Hawaiian-Americans
12. **HOWARD W. HALLMAN**
    Neighborhood Government in a Metropolitan Setting
13. **RICHARD J. GELLES**
    The Violent Home: A Study of Physical Aggression Between Husbands and Wives
14. **JERRY L. WEAVER**
    Conflict and Control in Health Care Administration
15. **GEBHARD LUDWIG SCHWEIGLER**
    National Consciousness in Divided Germany
16. **JAMES T. CAREY**
    Sociology and Public Affairs: The Chicago School

Planners and Local Politics

Impossible Dreams

Anthony James Catanese

Volume 7
SAGE LIBRARY OF
SOCIAL RESEARCH

SAGE PUBLICATIONS Beverly Hills London

For information address:

SAGE PUBLICATIONS, INC.
275 South Beverly Drive
Beverly Hills, California 90212

SAGE PUBLICATIONS LTD
St George's House / 44 Hatton Garden
London EC1N 8ER

Printed in the United States of America

International Standard Book Number 0-8039-0378-2(P)
0-8039-0397-9(C)
Library of Congress Catalog Card No. 73-94287

SECOND PRINTING

To My Wife, Sara

TABLE OF CONTENTS

PREFACE

This book is based upon certain beliefs in the righteousness of planners. It also is based upon belief in the basic and intrinsic honor of political leaders—even in post-Watergate American life.

If the book appears to be harsh, it is because of the urgent need for change that I advocate. I am not a journalist, and I have no intention of writing an exposé on planning and politics, nor do I desire to produce yet another academic treatise on the matter. I am a student of planning, and I want to raise basic questions about the relationships between planners and politicians.

During my university studies, I spent eight years examining the theoretical, methodological, and philosophical intricacies of planning. My orientation was toward such fields as management, systems analysis, and social science. While I am still convinced that there is much merit in planning, my professional experience in this field since 1962 has shown me that the integrating planning into the political process is equally as problematic as integrating it into life itself. My professional experience has been such that I can no longer accept theories that hold planning is distinct from the political actions needed for plan-effectuation. Plans that are formulated and merely given to politicians or other professionals for execution are doomed to dusty shelves.

What follows is a pragmatic series of discourses on planning and state and local politics which are reinforced by true case studies. I have chosen to deal primarily with state and local problems rather than with national and supra-state problems. My reason for doing so is that most planning is done at the state and local level, and the direct relationships between planners and politicians are more interesting at these levels. The case studies were derived from personal contacts with planners and politicians, as well as from my own experiences. I chose to use the generalist style not just to preserve anonymity, but to illustrate the almost universal nature of the problems. Furthermore, I believe that this approach to case studies can overcome the necessity to provide countless details of background for each analysis—enhancing readability.

Before beginning these discourses, it should be made very plain that there are a number of assumptions upon which this book is based. The major assumption is that there will be a continuation of both democratic and republican forms of government at the state and local level (which by extension would apply to the federal government as well). It is assumed further that state and local governments will remain as institutions which are limited in their activities through scarce resources. This implies that planning should be compatible with these assumptions, which means that planning should be practical and feasible. Planning that is practical and feasible ordinarily means planning that is not necessarily utopian and idealistic. If the reader does not agree with these assumptions, then there is a basis for contention at the outset. If, on the other hand, there is general agreement, a basis for exploration of complex matters is established.

To the many dedicated planners and politicians who have offered input, I express my gratitude to you in your anonymity. To colleagues past and present, I appreciate your advice.

I would like to acknowledge the dedicated assistance of Jane Wulkan, who edited, typed, and proofread the entire manuscript, and Maria E. Cuellar, who helped with the proofreading.

A.J.C.

PLANNERS AND LOCAL POLITICS
Impossible Dreams

Chapter I

QUIXOTE AND THE PRINCE:

INTRODUCTION TO PLANNERS AND POLITICS

> *"Where there is much wisdom there also is much sorrow"*
>
> —Buddha

> *"For I say unto you that except your righteousness shall exceed the righteousness of the scribes and Pharisees, ye shall in no case enter into the kingdom of Heaven."*
>
> —Jesus Christ
> Matthew V:20

Miguel de Cervantes (1547-1616) wrote of the decrepit old man who summoned up his last ounce of courage to beat the unbeatable foe. Ignorance, injustice, and power were to be replaced by wisdom, equality, and service. Don Quixote, assisted somewhat dubiously by the faithful Sancho Panza, was to rekindle the spirit of the knight-errant, of man in service to his fellowman with almost total disregard for his personal safety or for reward, for this is the classical definition of a knight. If Don Quixote seemed ineffective, easily beaten back–indeed irrelevant in today's jargon–it made little difference. Quixote and his faithful followers were not concerned with the petty realities of the times; they had nobler goals, impossible dreams.

Niccolo Machiavelli (1469-1527) observed the world around him. He wanted to describe his observations on how to win the most precious of man's victories: the struggle for power. Machiavelli's writings were primers for *Il Principe* on acquiring power, controlling territory, dealing with friends and foes, and compromising and forming coalitions. Machiavelli was no less concerned with the nobility of man than was Cervantes, but nobility was not to be an excuse for losing. *Il Principe* should be a realist about all and master the world in which he lived not just for domination but for order and rationality as constrained by man's ability. For Machiavelli believed that men need leaders who can enable wisdom, justice, and service to flower. Perhaps Machiavelli did not like the way things were, but he understood these conditions. Like others he may have wished for Utopia, but he had to deal with a real world first.

To some it may appear glib to offer an analogy of planners as Quixotic figures dealing with politicians as modern-day princes. Yet for purposes of general discourse, such an analogy seems more or less reasonable. Planners in America have been and are idealists attempting to lead crusades, with pathetic armies, against the cares and burdens of a seemingly unjust world. Too many times, the planner has been resemblant of the knight-errant, tilting at windmills, while the real struggles are won and lost without him. The politician fights for power, control, and coalition in the present. Planners still struggle to reach the impossible dreams for the future. At his best the planner is a local folk hero; at his worst he is an elitist and an irrelevant bureaucrat.

The discourses that follow have a simple objective: to suggest ways that planners in local politics can win battles. Some may find these discourses repugnant because they limit the spheres of planning concerns. Some may abhor the suggestions to deal primarily with critical problems that have some known probability of solution while deemphasizing others. Still others may find it curious that these discourses attempt to meld the quasi-scientific with the largely intuitive political process. But these are discourses on the realities of planning and politics, and how planning can work in a political world. And like Dante Alighieri this author must warn his gentle readers that voyages through an inferno are not pleasant excursions.

Roots of a Profession

To approach these matters it is necessary to take a critical view of the profession of city and regional planning, as well as the host of related professions and pseudo-professions that dabble in planning. Certainly the profession of planning has been castigated incessantly in recent years; indeed, planners have been more self-critical than have members of any other profession. Such a critical posture would be insufficient, however, if it were to omit the deeds and thoughts of politicians. The basic relationship has been long established: "The planner proposes, the politician disposes." Therein lies the crucial key to understanding the difficulties of planning: the planner has never functioned in a role wherein he had the ability to carry out his own plans. Planners must rely on the politicians to accept and implement recommendations of substance. The planner's role is part savant, part political eunuch.

As a profession planning was institutionalized in 1917 with

the founding of the American Institute of Planners (originally the American City Planning Institute) largely through the efforts of Frederick Law Olmstead and Flavel Shurtleff.[1] These two men, Olmstead the designer and Shurtleff the attorney-writer, epitomized the curious coalition of reformers and dreamers which organized a field of practice and study that became a profession. Yet the roots are tenuous. They are planted in a field of muckraking and utopianism which held that common people did not know what was best for themselves and that politicians were cunning devils bent on deception and corruption. The early planning movement grew out of reform activism coupled with a noblesse oblige of the concerned intellectuals and elite of the period. From its inception, planning as a profession was shielded from the evils of the political system and suckled by a benevolent aristocracy.

With growing institutionalization planning broke away from its elitist home and resettled in a bureaucratic maze. Protection against political involvement was guaranteed by the establishment of "independent planning commissions" designated as representative of the community and free from politics. While politicians were willing to play the independence game, they retained final decision-making authority as the only *elected* representatives of the people. Thus arose the American dichotomy of planning: planners had to be shielded by commissions and other devices from the elected representatives of the people. An entire body of knowledge, with accompanying techniques, emerged from this faulty and dated assumption. The result was predictable: more than a half-century of ineffectiveness in planning.

Localization of Planning

Planning as a governmental function has been recognized as a local responsibility since the City Beautiful Era led to the creation of Plan Commissions in Chicago, New York, and elsewhere in the first decade of the twentieth century. As the result of many factors, not the least of which was distrust of central government, planning remains a local power with enabling legislation emanating from states. There has never been a major movement toward centralized, national planning except for a brief flirtation by the National Resources Planning Board during Franklin Roosevelt's New Deal. Furthermore the only supra-state planning of major success and significance has been the work of the Tennessee Valley Authority, an entity that must be regarded as unique in the United States. This limitation on the governmental organization for planning is obviously an American trait and unlike planning in Europe, Scandinavia, Asia, and Latin America.

It can be argued that the fruits of planning in some countries have been of historical significance. Certainly the New Towns of Great Britain, Russia, Finland, Sweden, and Israel are historical. The major greenbelt areas around cities in England and France are of historical significance. Successes in land development in the Netherlands, France, Israel, Finland, and Sweden will be historical accomplishments. Social planning programs in Asia, Scandinavia, and Latin America, each within its given socioeconomic context, are historical events. Ironically the United States has emerged as the greatest contributor to the theory and technique of planning, yet has the worst record of success. If the premise that land use controls have been failures is accepted for the moment, it is not too difficult to see that there have been no historically significant accomplishments made by planners in this country.

Obscure and changing relationships between a free market economy and capitalist system and the responsibilities of government have been an impediment to planning successes. By its very nature a capitalist system necessitates a sophisticated planning process for the private sector. (Even the historical role of the federal government has required a strong process of planning for national defense.) Despite growing demands for more and better public services, however, our system of public sector planning for domestic matters has been weak. In its early stages public planning was viewed by many as one more step down the road to a socialist economy. The early planners contributed to this antagonism by their insistence that uncontrolled development of land and resources was inimical to the public well-being. The people attracted to the planning profession have been liberal in general in their attitudes toward the role of government and dubious of a free market that had no constraints. The basic question, then, has not been resolved, and some unseen hand limits the areas of concern for planning endeavors by the national government.

One overall result is that the practice of planning in local government takes three general forms. The Participatory Form holds that planners must seek maximum involvement of people in their communities in order to formulate a basis for political pressure to effectuate plans. The Technocratic Form dwells on the premise that planning problems are basically technical matters requiring an institutionalized expertise. The newest form is the emerging Activist Form which presupposes that planners can best accomplish their ends by serving as politicized aides to elected leaders. Of these three general forms of planning practice, only the last has strong potential for effectuation of planning proposals, although such planning has tended to be short-term and specialized in nature. The second form, however, probably accounts for the majority of practicing planners.

Unlike in other countries the planner in the United States has evolved as an institutionalized technician who dabbles mainly at the periphery of the political process. His security and comfort are guaranteed by such a role, and serious problems have arisen only when he tries to overcome political considerations on technical grounds.[2] Rarely is the comfort and security of the planner threatened by his inability to have his plans fulfilled. In this sense the planning profession can survive for many years in its institutionalized format. It does not mean, however, that more critical areas of planning will not be assumed by others less concerned with longevity in position and more concerned with politically sensitive matters.

The major problem that has challenged the planning profession in recent years is that it has been the scapegoat of political failure, a stereotype that has led to a "damned-if-you-do, damned-if-you-don't" schizophrenia for planners. The institutionalized and low visibility profile exudes an aura of elitism and ivory tower boogeymen. Thus a southern civil rights leader can proclaim loudly that "planners have been the blame for racial segregation in this country." A midwestern governor announces that "planners at the state and national level have imposed social planning disasters that are destroying the fabric of American Society." A middle-class engineer in California running for state office campaigns on "controlling the planners who have been destroying our environment." Incredibly a profession that criticizes itself for ineffectiveness is developing a political image of great power which can be unleashed for creating all forms of societal evil. At the local level when something goes wrong because of the lack of planning success, the planner serves as a latent target for blame because of his "bad planning." Certainly the polity realizes that developers may be greedy, and politicians may be wrong at times, but it appears that the planner is the one who is screwing up everything.

A stereotype of planning as a localized public function thus emerges. It is localized because, as mentioned, national planning has never been popular in the United States. Local politics are assumed to be something that the planner must avoid, and he must deal with supposedly nonpolitical bodies rather than elected leaders in most cases. The local planner is ordinarily a technician who argues against political considerations on the basis of supposedly objective analysis. Such a stereotype forces an isolation of the planner from community and special interest groups, and he appears as an alien overseer to many. The expected result of this stereotype is a growing antagonism toward planners which may not jeopardize their existence but may force changes in certain aspects of the planning process.

The Federal Strings

While planning is primarily a local function, it is not unnoticed by the federal government. The influence of the federal government on local planning is indeed great even if indirect by nature. Federal influence arose largely through grant programs, the most important of which was the Comprehensive Planning Assistance Program originally authorized by Congress in 1954. Through this program, usually referred to as the "701 program" (because of its numerical title), as well as planning grants for transportation, housing, open space, law enforcement, and social services, the federal government has funded a great deal of state and local planning.

Even with the advent of general and special revenue-sharing, the federal government does not give money to local governments without strings attached. Usually these strings are "guidelines" for or to be eligible for the funds and must satisfy other guidelines for acceptable completion of a planning project and additional funding. Hence, without any direct political role, the federal government through its bureaucracy has profound controls and powers over local planning.

The general nature of the guidelines for planning funds has been nonpolitical. They usually entail a listing of essential elements to be included in the planning and technical guidelines for undertaking such projects. Except for some dubious guidelines that required citizen participation and involvement in the planning process, many of which were incredibly vague and time-consuming, these guidelines have made it more difficult to relate planning to state and local political processes and have contributed to the further institutionalization and technocratization planning.[3] The federal funding programs have provided local planners with a justification for spending more time on data collection and analysis than on grass roots planning involvement. Furthermore the federal planning funds have given birth to a formidable and unwieldy new growth industry for not only planners but all sorts of professionals and pseudo-professionals seeking new, interesting, and lucrative areas for study and learning which might even result in problem-solving. As a former director of the planners' professional society said, "Life in these United States has gone into limbo, and effectively stopped, until we complete our studies and propose new studies for improving the quality of life."

Much of the rhetoric concerning revenue-sharing was based on the need to decentralize the federal government and return decision-making power to local government. This was indeed heresy and anathema to many liberal congressmen and especially the federal bureaucracy. Nonetheless revenue-sharing is fact, and the rationale for it remains in theory. In reality the federal guidelines remain although they have been generalized and are not quite so detached. Guidelines for planning have been broadened so as to include an emphasis on many everyday problems of government. While it is early to evaluate the effects of these recent changes in federal policy, there is little to indicate that major changes in the planning process are forming as a result. The more likely effect will be that revenue-sharing and planning and management grants will remove some of the previous barriers to planning effectiveness in the local political arena. The basic battles will still have to be fought and won by planners in that arena, however, and it will be slightly more difficult to justify failures as being caused by unrealistic guidelines.

A benefit of the federal role in local planning has been to compel adherence to what may be "sound" principles of planning. In the earliest stages this meant that local governments were required to have master plans covering land use, transportation, and housing, at the minimum, with commitments to achieving these plans through such basic tools as land use controls (zoning, official maps, and subdivision controls) and capital budgets. Hence local politicians were coerced gently into making commitments, albeit tenuous. Extensions of this concept were applied to environmental impact, open housing, regional government, and federal program review powers over state and substate regional planning agencies. As planning has existed in this country it is doubtful that planners could have introduced these principles into local politics without the pressure and power of federal support. There are still many who argue that state and local government cannot uniformly incorporate such planning principles without federal intervention, and hence, they challenge the rhetoric of revenue-sharing and related federal programs.

The inherent fallacy in the argument for strong federal guidance of state and local planning is the assumption that a uniform minimum level of planning is better than diverse levels. Several states and cities have planning programs that are superior to any federal minimum program. Most states and local governments have planning and political processes which are unique. It is comfortable to generalize these differences and talk of urban needs and state needs, yet the underlying differences cannot be ignored.

A minimum planning program or model planning program which is designed by the federal government is sheer folly. The benefits of bringing states and cities with poor planning programs to some national minimum do not outweigh the costs of slowing down the more progressive states and cities. Furthermore the intricacies and subtle complexities of state and local politics make model planning programs and organizations an absurdity. What appears to the outsider as a poorly programmed and organized planning function may be the best possible arrangement for effectiveness within the given situation. Conversely the attempt to instill a sophisticated planning technocracy in a political structure composed of "good ol' boys" and "wool hats" would seem bound to guarantee an extrapolation of past ineffectiveness of planning.

Dilemmas of Politics and Planning

Ours is a pluralist society which exists and survives within a context and culture of different and particular interests competing for political power and benefit. There are those who argue that America is no longer pluralist and has evolved into a mass society with a monolithic set of interests; yet this argument is negated by simply taking a stroll in an urban area, or attending a city council meeting, or reading the newspaper, or participating in any way in the vibrant and exciting reality of politics.

The basic unit that functions to represent the pluralism of American society within the political process is the *special interest group.* Certainly individuals have considerable power, and certainly individuals exert special interests, but it is the organized special interest group that occupies the paramount place for representation. The degrees to which groups may be organized, their numbers, their resources, and their power will obviously vary immensely. The special interest group may be the General Motors Corporation or the Druid Hills Civic Association of Atlanta; the National Association for the Advancement of Colored People or the Ku Klux Klan; or the American Institute of Planners or the Square Earth Society. The special interest group should be viewed in its broadest perspective in order to appreciate its role in politics and planning.

Prior to delving into specific lessons to be learned from actual though anonymous case studies involving special interest groups, a number of hypotheses can be set forth. The basic hypothesis is referred to as the "Catanese Contention."

> *The local political process usually will overrule long-range and comprehensive plans based solely upon rationality principles of planning.*

The local political process can be defined as that set of actions and reactions settled by elected officials responsible to a polity. The polity is represented by key individuals and special interest groups. Rationality principles of planning are generally those which seek "the greatest benefit for the greatest number" and will ordinarily conflict with key individual and special interest groups unless modified especially by the planners. Rationality principles are inculcated in techniques and methods of planning such that concepts of cost-benefit and optimization are the primary output of technical endeavors. The Catanese Contention hypothesizes that such a practice of planning, while admirable in an intellectual arena, is too inflexible for a political arena. The politician must weigh supports and demands for given programs and the relative equity of those competing forces. The planners seek to form an argument that refutes special interest groups in favor of benefits to the elusive public interest.

The dilemma is that the planners' recommendations ordinarily are based upon different decision criteria than those of the politicians, and hence the politician will make decisions based upon political power and equity rather than technical merits. This dilemma does not preclude the politician from supporting planning or from endorsing very generalized plans. It does hypothesize that substantive and specific decisions, however, need not be effected by planners' recommendations.

There are several axioms to the Catanese Contention.

Axiom I. Special interest groups often outweigh general public interests. The crux of American democracy at the state and local level is that special interest groups are guaranteed an equal footing with general public interests. The political reality is that general public interests are almost impossible to define or articulate. On the other hand special interests are not too difficult to define and articulate. The planners attempt to formulate goals and objectives for the general public through largely technical methods. Special interest groups formulate their demands on the basis of their own objectives and their equity in the political system, and they attempt to demonstrate that their demands are really good for the general public, or at least will do it no harm. Because the special interest groups are adept in the political arena, especially when they possess sufficient economic muscle, the planner is faced with an insurmountable task of arguing that a nebulous public equity is the greater. The game is not a fair one for the planner because he has nothing to gain or lose other than the virtue or rationality of his cause. The equity of special interest groups has definable limits, however, and a win and lose calculus can be estimated. Ordinarily the planner can be expected to lose the game unless unusual or intervening variables are introduced.

Axiom II. Special interest groups can be more effective through overt political actions than through involvement in the planning process. Special interest groups have found that actions such as political support, campaigning, protest, and reaction can elicit responses from elected officials. Through persistence and political activism they can eventually succeed to some degree. Attempts by planners to involve such groups in a technical planning process have not demonstrated an equal chance of success. A basic problem is that special interest group participation in the planning process means essentially a removal by at least one step from direct communication with the elected decision-makers. Furthermore it can dilute the political power of group leaders since the planner often emerges as the advocate for the group. Few groups can afford to risk such a dilution of their power and prefer to deal more directly with politicians. Hence the potential client of the planner is dubious about this role as the articulator and advocate. While special interest groups often cooperate with planners and seek their support, they want it clearly understood that only they speak for themselves and will abandon their cooperation readily when their special interests are undermined. A disturbing spinoff of this axiom is that the planner is sometimes forced to invent a general public interest when he cannot coalesce special interest groups around an issue with a rational planning solution.

Axiom III. The planning process can be modified so as to be effective under the conditions of the Catanese Contention. In essence this axiom means that planning and effective implementation through the local political process is possible. Major restructuring of the planning process is required, however, in order to recast the planner more in the Activist Form with substantial inputs of reasonable scientific techniques. The planner as a politicized professional would necessitate a role other than that of the institutionalized spokesman for the general public or the Technocratic Form. It would require also elements of risk-taking, dealing with political equity, and respect for the autonomy and power of special interest groups.

Windmills and Religion: An Analogy

A return to the Quixote and Principe analogy is interesting to conclude this introduction. The analogy deals with the essence of the political process: power and honor. Both Cervantes and Machiavelli wrote of the occupation and control over Spain by the Moors. The political questions surrounding Moorish power offered fascinating commentary on political effectiveness.

Cervantes' work was largely a parody on the romantic imagery of knighthood and war. The principle actor is of course Don Quixote of La Mancha, a fifty-year-old gentleman who was obsessed with the rise of evil in Spain and the dying of the knight-errant legend. Don Quixote would set out to change this by combatting evil and raising the worship of his quest and purpose. He was to be a knight in service to mankind seeking to reach the impossible dream of a just and peaceful world and in the process attain the reknown and honor he felt he deserved.

Don Quixote was no idle dreamer. In his own way he was like a planner. He had educated himself through reading voluminous accounts of early knights. He had surrounded himself with the weapons and accoutrements of knighthood and had practiced combat daily. His life was regimented and ordered, and he tried to incorporate his beliefs in the Divinity in all that he undertook.

The first order of his adventures would be to right the wrongs and abuses perpetrated by the Moslems in Spain. Nobody had encouraged Quixote to begin his quest, and the local prelates of La Mancha were skeptical and ridiculed him. Yet the Don felt a higher calling to humanity and embarked upon his first adventure.

The first travels of Don Quixote and his squire Sancho Panza were beset by a series of minor rebukes and setbacks. In order to purge Quixote's soul, the local curate and barber searched his library for the cause of his obsession and burned his books on chivalry. After a brief return Quixote set upon his second travel even more determined.

Outside Malaga, Don Quixote spied some thirty or forty monsters, probably of Moorish design, which were guardians of evil and wealth. Quixote spoke of "the great service unto God to take away so bad a seed from the face of the earth."[4]

"What giants?" asked Sancho Panza. "I pray you to understand that those which appear there are no giants, but windmills; and that which seems in them to be arms, are their rails that, swung by the wind, do also make the mill go."

"It seems well," spoke Don Quixote, "that thou are not yet acquainted with the matter of adventures. They are giants, and if thou beest afraid, go aside and pray, whilst I enter into cruel and unequal battle with them."

Despite Sancho Panza's cries, Don Quixote spurred his horse, Rozinante, and raised lance. As he charged the wind picked up gently and spun the rails, which Don Quixote interpreted as an act of cowardice and flight.

"Fly not, ye cowards and vile creatures. For it is only one knight that assaults you," Quixote shrieked.

As the lance hit the rail, the weapon splintered. Both Don Quixote and Rozinante were plucked from the earth and thrown very far into the field out of reach of the evil windmills.

As Sancho Panza approached them to offer aid, Don Quixote intoned: "Peace, Sancho, for matters of struggle are more subject than any other thing to continual change; how much more, seeing I do verily persuade myself, that the wise Frestran, who robbed my study and books, both transformed these giants into mills, to deprive me of the victory, such is the enmity he bears toward me. But yet, in fine, all his bad acts shall but little avail against the goodness of my lance."

Don Quixote had lost his lance, however. He recalled reading about another great knight, Diego Peres, who lost his lance in a battle with the Moors. To continue the battle he cut a large branch from an oak tree, and he went back into the fight and battered many Moors. Don Quixote saw an oak tree and searched for a suitable branch so that he could continue his pursuit of honor and righteousness.

Machiavelli believed that only great objectives and personal successes could bring honor and power to a leader. The King of Spain, Ferdinand of Aragon, was considered most exemplary of this principle. Ferdinand was a weak king at first because of the power of the Moslem forces in Spain, as well as the large number of small fiefdoms that prevailed among Spanish nobility. King Ferdinand was a planner, however, who sought to unify Spain and cast out Moslem invaders—and perhaps even more.

The dream had to be supported and honored by a strong political body because it was not sufficient or likely that an army could be raised solely on the basis of the cause. Such a political force was obviously the Catholic Church. The church supported and blessed Ferdinand's quest. He raised an army that defeated the Moors at Granada and the Barons of Castile. The movement grew and the peasants came to both support and cherish the King. Feeling that the church had established the base of his power, Ferdinand started the expulsion of the Moors from Spain which was completed by 1610.

King Ferdinand felt that he owed great loyalty to the Catholic Church because of its role in his rise to honor and power. He also felt that it was wise to pursue his plans in the name of protecting the Catholic faith. He attacked Africa, France, and Italy on this premise and won major victories.

In Machiavelli's words: "He has always planned and completed great projects, which have always kept his subjects in a state of suspense and wonder, and intent on their outcome. And his moves have followed closely upon one another in such a way that he has never allowed time and opportunity in between times for people to plot quietly against him."[5] Machiavelli was able to generalize a political principle from his analysis of King Ferdinand: "A prince wins prestige for being a true friend or a true enemy, that is, for revealing himself without any reservation in favor of one side or another. This policy is always more advantageous than neutrality."

Are there lessons of analogy that can be drawn for planners and local politics? At least there are some insights available. The two situations described above have components needed for planning and execution of plans: objectives, strategies, and support. Ironically the objectives of Cervantes' and Machiavelli's heroes were essentially the same. Both had arrived at these objectives in part through personal ambition and in part by assimilating what appeared to be popularly held desires and needs. Both Don Quixote and Ferdinand had reached their conclusions as to what objectives were valid through a process of fusing their own values with perceptions of public values: it seems futile to assume that such objectives can be set without this sort of value judgment.

A major difference between the two characters was Quixote's isolation from any special interest groups and his preferring to deal with a more nebulous mankind. Ferdinand sought more specific support through powerful groups. The latter was more successful because he had strong backing, while the former found that the representatives of the church and other institutional powers came to oppose him. A possible lesson is that special interest groups can be most useful and necessary in attaining objectives. On the other hand, attempting major accomplishments without any specific support while relying on the logic, honor, and support of society in general (undefined) does not seem to offer promise of success.

The result of such differences in objectives and supporters is that strategies will be different. Cervantes' character had little strategy because he was relatively unconstrained; he could attempt anything he wanted. Ferdinand was constrained by both his objectives and his support group interests. He had to carefully plot his actions so that specific objectives could be reached and special interests kept relatively satisfied. Such a strategy was much more active and energetic than the flip and unfettered strategies of Don Quixote. Formulation of strategies or alternatives to strategy is more difficult when such constraints exist than when they do not. The unconstrained strategy or plan is the easiest to develop because it must be acceptable only to the planner.

Formulation of plans or strategies to attain plans without constraints and supports from interested groups can easily become a worthless exercise. This is plausible even in the event that the objectives to which the planning is addressed are generally accepted or sufficiently vague so as not to upset anyone. Why then would any planner undertake such an approach? Like the Quixote analogy it may be that some planners are convinced they are right and the rest of mankind must only realize this before acceptance. Reason will triumph over unreason. An alternate explanation and analogy would be that like Quixote, perhaps, some planners do not really believe that they can change very much in the world and are content to play the roles of crusaders and tilt at windmills even if only the appearance of a brave defender of the public is garnered.

And Penguins

The planning department of a large midwestern city recently adopted the penguin as its mascot.[6] The unanimous vote was based on the fact that the penguin also was a "strange bird." A resolution was drafted which further explained the reasoning for the move. The resolution noted that while the penguin was "esthetically pleasing in design, it could never fly." Penguins, like planners, the resolution goes on, "have an elegant stance, yet face all decisions with cold feet and huddle together in groups." Finally, the resolution was intoned, "penguins have few friends except those of their own kind." The recommendation and resolution arose from the professional staff and were sent to the nine-member independent policy commission for approval. Only one commissioner expressed reservations. He said there are four kinds of penguins–the King, Emperor, Adelie, and Jackass Penguin

–and he wanted the public to be certain which kind of penguin was being proposed. He later removed his mild protestation and voted affirmatively after the city attorney offered his opinion that the public would be well aware as to the kind of penguin the planners wanted as mascot. The professional staff was pleased with the action and suggested that other planning departments adopt suitable mascots.

NOTES

1. For an excellent review of the American city planning profession from 1890, see Mel Scott, *American City Planning Since 1890* (Berkeley: University of California, 1969).

2. An interesting case study of this point is reported in D. R. Judd and R. E. Mendelson, *The Politics of Urban Planning: The East St. Louis Experience* (Urbana: University of Illinois Press, 1972).

3. For a specific example, see Daniel P. Moynihan, *Maximum Feasible Misunderstanding* (New York: Free Press, 1969).

4. From *Don Quixote of the Mancha,* Book I, Chap. VIII.

5. From *The Prince,* Chap. XXI.

6. From *Planning: The ASPO Magazine,* Vol. 38, No. 8 (September, 1972), p. 210.

Chapter II

ON THE DIFFERENCES BETWEEN PLANNING AND POLITICS

> *"The sublime and the ridiculous are so often so nearly related, that it is difficult to class them separately. One step above the sublime makes the ridiculous, and one step above the ridiculous makes the sublime again."*
>
> –Thomas Paine
> *Age of Reason*

The story is told by a young legislator of the planner who had worked many years in a large southern city. The legislator relates that the planner had developed a master plan for the city and had been attempting, somewhat fruitlessly, to implement the document. After years of failure the planner learned that he had a terminal disease and would soon expire.

"But I have so much left to do," he told his physician.

"I am sorry," the doctor replied.

The planner passed away and gained consciousness in an ethereal and strangely familiar place. Why the place was so familiar was unclear, but he thought it to be the loveliest and most charming city he had ever seen. He realized that this perfection was indeed the materialization of his master plan for his earthly city.

As he walked through this Utopia, he saw an ominous figure of obvious authority.

Approaching the figure he said, "Sir, I had no idea that perfect plans could reach such a high level of implementation here in heaven."

Somewhat puzzled and angry, the looming figure spat back, "That is not where you are."

How does one determine what is good planning and what is not? Such a question has plagued practitioners of the planning profession since its inception, yet complete answers have never been determined. To some extent plans for the future are dreams cast in either pragmatic or visionary perspectives, but some dreams can be nightmares. It can be far more oppressive and destructive to a society to construct environments that are unwanted than to do nothing. Despite the old maxims, bad planning and bad political leadership are worse than no planning and no leadership.

Politics may be hell, but the planner is in it with the politician. For various reasons the planner may sometimes refute this point and argue that he is merely touring the area like Dante. Some planners may believe that they really are not in the place and simply send messages there for action by the true inhabitants. Other planners do not know that politics, politicians, or hell exist, and live their lives out quietly and comfortably having little or no effect upon the rest of the universe.

Ramifications of this set of conditions include rather obvious differences in the way that planning is undertaken and the way that politics take place. Both planning and politics occur more or less as a process, that is, a series of actions that are conducive to an end. While there are dynamic aspects of these actions, it becomes clear that there are distinct processes for planning and politics. From place to place, time to time, and context to context, it appears that the planning and political processes are predictable to some degree and tend to have areas of considerable difference and misfit.

Rudiments of the Political Process

There are many interpretations of the manner in which the political process does, could, and should operate at the local level. This field of study has been one of the most exhaustively tilled in all of political science and related disciplines yet one of the most difficult to understand. One of the more interesting approaches to describing the political process has been to consider it as a *system* or a whole composed of interdependent parts. All of the parts must work together to create the whole, and the whole in fact becomes somewhat more than the mere sum of the parts.

A basic theory that has been applied to the way that systems work has been the elementary theory of the field of cybernetics which is the study of control and communications between man and machine.[1] This basic theory, often called the General Cybernetics Model, holds that all systems operate through a conversion mechanism which transposes certain inputs into outputs. This conversion occurs according to structured and orderly rules and regulations which may vary from time to time but are largely known. This is similar to saying that certain rules of a game exist that may lead to varying results.

A central concept of the General Cybernetics Model is the so-called feedback loop. In its most elementary form this feedback loop means that the users or controllers of the system have certain reactions to the actions or output of the system. This feedback can be positive, negative, or neutral to various degrees of severity and complexity. In machine systems feedback can be automated through self-regulating devices so that the system is adjusted to meet the needs expressed through some set of standards. For example if a heating system is set at 70 degrees in a home, and the thermostat indicates that the room temperature is above that, the feedback loop will cause the heating system to reduce its work in order to allow the desired room temperature to be attained. In social systems which are not automated, feedback takes more complicated forms of action. Examples of feedback loops in politics might be protests, disruptions, direct intervention, support, satisfaction, or apathy.

The General Cybernetics Model, shown in Figure 1, of input, conversion mechanism, and output which is subject to feedback usually does not operate in a vacuum. Whether the system be mechanical or human, it must function within an

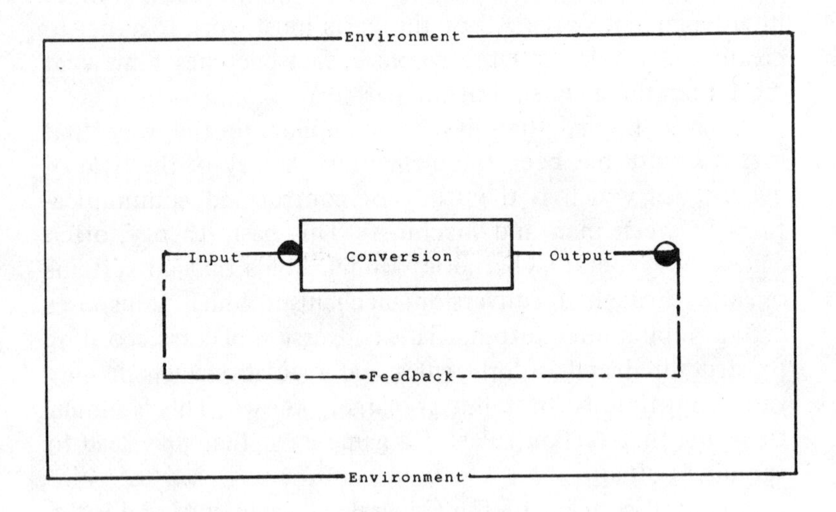

Figure 1. THE GENERAL CYBERNETICS MODEL

environment. The environment can establish both the limitations and potentials for the system. In mechanical systems, temperature, humidity, climate, pressure, and other conditions of the physical environment can set both the constraints and effectiveness levels. In nonmechanical systems such factors as desires, needs, wants, morals, and ethics, which are generated by the nonphysical environment, can establish limits and potentials with equal force. The environment thus emerges as a critical consideration in the General Cybernetics Model because it has great effects upon the system yet is itself a dynamic and elusive entity.

There has been much debate in recent years as to whether or not such a model or way of examining systems can be useful for political life. There seems to be ample justification for using such an approach, however, if the analogy is used for purposes of generalized explanation and not necessarily for operationalization. This means that we can use the General Cybernetics Model to explain the way that the political process works in general, but we can get somewhat foolish if we attempt to build computer programs or machines that serve as surrogates for politics and politicians. The General Cybernetics Model in this perspective offers considerable promise for understanding politics, as well as how the planning process interacts with the political process.[2]

All of this assumes that the political life of local government is in fact a system or at least system-like. Is this a reasonable assumption? One of the foremost students of this problem is David A. Easton who argues that the analogy is valid for study and analysis since politics is the allocation of power and values which result from interactions among different parts of society.[3] Many political scientists and observers of the political scene find this approach of great interest and utility.

Easton has constructed a model of the political process (see Figure 2) which borrows from the General Cybernetic Model. The input to the conversion mechanism is focused on

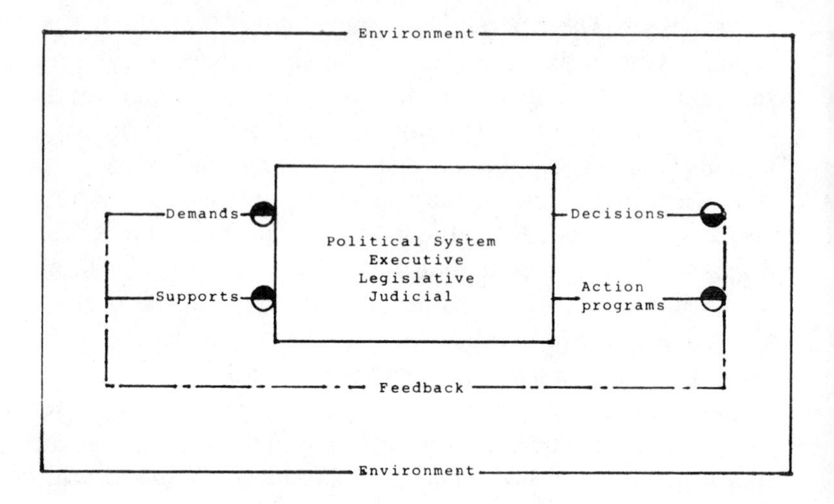

Figure 2. THE POLITICAL PROCESS

two major political variables, supports and demands. Demands for the conversion mechanism, which is represented by government and its executive, legislative, and judicial branches, are generated by special interest groups and key individuals. It has been speculated that in the real world of local politics, these demands are often articulated and detailed through institutionalized agencies including the planning offices. Many of the demands for new facilities and services, for example, may be initiated by special interest groups but usually emanate formally from a public agency, if the system is working well. For every demand or set of demands, there is a corresponding set of supports which are positive, negative, or neutral in varying degrees. These are fixed also by special interest groups and key individuals and tend to have attachments to the relevant offices of the bureaucracy. One aspect of political systems that Easton believes to be unique is what he calls a *withinput.* This is a demand and support generated by participants in government who may be involved directly in the conversion or decision-making. More than simple porkbarrel projects are involved because often the withinput is linked to demands and supports that have not been articulated but have been derived through intuitive and implicit approaches to political rationality. This often leads to a chasm of misunderstanding between the technocrats and politicians.

The conversion mechanism receives the demands and supports and will act upon them if there are sufficient questions of power and value. When the demands and supports are inarticulate and abstract, or when the consequences are particularly unknown or dangerous to the political system, the conversion mechanism may not act upon the input. Aside from this inaction, the conversion mechanism, which is composed basically of the elected members of the three branches of government, generates decisions and action programs. Decisions are responses to the demands and supports. Action programs are commitments usually through the allocation of resources to satisfy the decisions. As can be readily noted in complex instances, the conversion mechanism may render a decision which is responsive to the demands and supports but for any number of reasons it may not provide the resources necessary for effectuating the decision. When the system is working ideally, the programs are directly related to the level of commitment implied by the decision.

The feedback loop exists in the political system in order to provide both stability and change. This is possible because a political system with no feedback loop can produce decisions and programs that are so obnoxious to the society that it leads to overthrow and replacement of the system. At the same time the feedback loop allows the decision-makers in government to adjust their thinking and often change their minds. The feedback loop usually works by sending the decisions and programs back to the special interest groups and key individuals for additional and revised analysis. The input is then recycled through the conversion mechanism resulting in perhaps a different or modified set of decisions and programs. It is possible also that the feedback loop could result in changes being made in the structure of the conversion mechanism itself or even changes in the individuals holding positions in the conversion mechanism. The latter is usually a long-term effect of feedback and is rarely a direct result in all but the most crucial issues.

The environment of the political system is most complex. The environment is both the set of interactions based upon values, beliefs, and wants of people at the state and local level and the general atmosphere established by national, international, and societal values, beliefs, and wants. The environment for local political systems is composed also of a set of conditions that emanate from the economic, legal, and social forces that are temporal and evolving. For this reason it is incorrect to assume that the same set of inputs will result in similar output consistently over a period of time, and the effective and astute participant in or observer of the political system must be aware of the critical nature of timing.

The major lesson that can be learned from this analogy of systems, cybernetics, and politics is perhaps intuitively obvious. The politician is cast in a role of decision-maker even though he is affected by all sorts of influences and plays a game with fairly clear rules by which he must weigh the demands against a series of supports and then issue responses and concomitant means for reaching the desired results. In this sense politics is certainly a rational process if one accepts the political system as defined. The political system is inherently honest and effective because personal gains by the participants in the conversion mechanism tend to violate the rules of the game. The political system basically is stable yet has attributes that allow for change and dynamism through both the environmental impact and feedback effects. All of this is possible when the political system is working well at the local level. When it is not working well, the need to analyze the system for maleficence and inefficiency becomes the responsibility of the society that created the system.

The planner can take several roles in the process as defined. The traditional role of the planner, referred to as the Technocratic Form previously, has been to act as the overall adviser and reference point for the elected persons responsible for making decisions and resource commitments. The planner has functioned in the political process as a didactic Nestor or Egeria in this classical guise. The planner as an actor in the political process in this fashion would rarely take a strong position because his offering to the process is objective and technical wisdom free of the influences from the demand and support groups. Such a role is highly prized and cherished by many intellectually inclined professionals and hence has been fitfully defended by the planning profession.

A somewhat rebellious group of younger planners in recent years has argued that such a role, while glamourous, is ineffective because it deals with all those matters that have little bearing on the decision (recall the Catanese Contention). A number of leaders of this group of planners called for direct advocacy of the demands and supports of given special interest groups so that the forum for decision-making had a more reasonable basis for weighing the input. This planning approach, usually called Advocacy Planning, has had a mixed success and incredibly strong reaction against it.[4] It is analogous to the Participatory Form mentioned previously. Advocate planners insist on working within the political system like lawyers for the clients of the system. Much of the movement was decidedly radical and perhaps naive, which resulted in an almost direct relationship between the advocate role and the assistance offered to those special interest groups which had the least ability to formulate their demands and supports in a clear manner. This usually meant ghetto dwellers, welfare mothers, militants, minority subgroups, and elderly groups, the special interests with the least political power base for submitting input to the political system.

The Activist Form has been found among some planners

for several years but has been relatively quiet and unheralded. Rather than being analogous to Nestor, this group sees a more proper analogy with Metternich as it seeks to exercise the personal and covert diplomacy necessary for reaching the decisions and carrying out the programs believed vital. Like the advocate planners, members of this group realize that the value-neutral role of the planner in the political process was limited and effacing. Both the Activist and Participatory Forms employ the risk-taking and tenacious actions that are effective in the political process as described.

While there may be instances where all three forms are used by planners to carve out a role in the political process, the predominant approach is that of the technocrat. The planner as the value-neutral, objective, and disinterested technocrat has been institutionalized and protected by the professional groups. Antipolitics laws for civil servants have afforded further shielding of the planner from deeper involvement with the conversion mechanism of the political process. As such the planner has a role in the political process which is often no more or less than any other group with no political power or value that is seeking to attain certain allocations of power and value from the elected decision-makers.

Rudiments of the Planning Process

With the role of the planner being relatively mild in the decision-making apparatus of the political system, but nonetheless required and institutionalized, it became apparent that a process for doing planning was essential. The planner had to determine which points of intervention in the political process were open to him and then offer the most sophisticated recommendations possible. Such a strategy implied that the greatest good for the greatest number and other Pareto-like conditions should be applied. It also meant that the planning process had to be reasonably scientific because the utility of the planner in his institutionalized role was largely his wisdom and objectivity. This approach was vital in the early years because politicians needed to be certain that the source of their advice was beyond question in terms of loyalty, which could only be guaranteed by employing planners who were apparently dedicated to the esoteric principles and practices of the profession.[5] Translated into stereotypes, this meant that the intellectual, highly educated, and visionary scholar-planner was one to be trusted but not taken too seriously. Such a condition was acceptable to most planners because it meant that little was expected of them and there was no true accountability involved. Many interesting problems could be studied and ego-satisfying solutions proposed, which resulted in suitable publicity and exposure to the community and, more important, to other planners. Like the penguin, the planner considers it very important to be a valuable member of his flock.

With so many constraints placed upon its scope of activities, the profession had to develop a way of doing planning that was optimal, in this sense the best that can be attained within the limitations imposed. The search for such a planning process has continued to be an obsession with theorist, student, and practitioner of planning for more than a half century in the United States. There was no hesitancy to forage into other professions for their approaches and tools, and hence planning for the public sector became a synthesis of many fields of study.

Coleman Woodbury offers probably the best description of what has evolved by describing planning as the "process of preparing in advance, and in a reasonably systematic fashion, recommendations for programs and courses of action to attain the common objectives of the community."[6] Planning is done in advance because only the future can be planned: the present and the past cannot be planned. It is debatable as to how far ahead we should plan, but it is more or less agreed that there is a near-term and long-term future which can be planned to some degree. Planning is reasonably systematic which has meant scientific to most planners. Planners have had a fascination with science because of its clean and sharp results and have tried for years to incorporate its techniques within their own. Planners recommend programs and courses of action to decision-makers but almost never function in anything more than an advisory capacity. This seems in keeping not only with the limitations placed upon planning by the politicians but with some firmly established beliefs that only persons who have been elected to office should be allowed to make decisions in local government. Lastly Woodbury notes that planners are concerned with common objectives of the entire community rather than with any special interests, individuals, or themselves. While it is somewhat unworkable, this last tenet is necessary if planning is to have any kind of theoretical basis that can utilize scientific concepts of rationality.

A process of planning has evolved that is related to the Woodbury definition of the field. While there could be argument over specific details, the following generalized description of the process would probably be acceptable to most planners. The planning process has seven basic steps:

(1) The identification and definition of problems and their interrelationships.

(2) The determination of the community's objectives in dealing with each problem, as well as the totality of problems.

(3) An appraisal of the existing plans, programs, and decisions for dealing with the problems to determine if they are adequate.

(4) The formulation of alternative recommendations for solving the problems according to the community objectives.

(5) The evaluation of the alternatives by analyzing the predictable by-products and side effects, as well as the approximate costs and benefits.

(6) A recommendation for adoption of the most appropriate alternative by the decision-makers.

(7) The possible modification of the recommended alternative according to feedback from the community and elected officials.

The process thus emerges as an effort to solve interrelated problems of a community as long as the community can espouse what it wants done. Using this process the planner can function more or less scientifically in drawing up the available strategies and proposing the best approach. Modifications are permitted when reactions indicate errors in determining community objectives for the problems, assuming that reasonable techniques have been used for assessment.

At the state and local level, problems that are suitable for inclusion in the planning process become numerous. Land use and development, transportation, community facilities, open space and parks, health facilities, social services, education, pollution controls and treatment, and criminal justice problems are only the major topics which exhibit fairly obvious interrelationships. Not only does the planning process require that the planner understand all these problems and interactions, but it requires that he be able to integrate the various solutions into a meaningful and coordinated strategy that is consistent with the community will. Such a process requires a substantial amount of time and effort if undertaken with even a modicum of seriousness. When placed in the perspective of the political system, however, such a process has similar weight to less substantive articulation of recommendations by special interest groups.

The basic problems with the planning process are that it promises to do too much, takes too long, covers too many problems, is too complicated, and produces results that are largely vague and esoteric. Some of these problems are directly related to the inbred problems of bureaucracy, as well as to the professionalized make-work attitudes of many prominent leaders of the profession. Some can be attributed to a desire among the planners to feel that they are contributing to the community's betterment by working hard to uncover the right answers to the problems, even though there may be some debate as to whether the right questions have been asked. Nonetheless, in the end, the planner may be somewhat ineffective in getting his recommendations accepted, but he can never be faulted for being lazy. The planning process necessitates extraordinary levels of work and activity and may have real value if for no other reason than showing that someone is busy trying to find the answers to all of the community's problems; there is some comfort in this thought.

The River Case

A case study illustrating the differences between the planning and political processes, as well as the respective roles of planners and politicians in the political process, can be drawn from the southeastern United States. A large river winding through four states passes through one of the larger metropolitan areas of the south. The river is relatively clean and especially so closer to its source in the mountains. Land abutting the river is essentially undeveloped, although it has been in the hands of several large estates for many years; therefore, the river has become a favorite recreation spot for canoeists, hikers, and nature lovers.

In recent years there has arisen considerable market potential for land along the river. Competent and shrewd developers bought up the land from the large estates and farms at very low prices, although the market tended to rise as word was spread about the boom along the river. In a short period of time, much of the land along the river that was close to the metropolitan area had been bought by speculators and developers, and only a few of the original landowners held onto their land either for their own enjoyment or for yet higher prices.

Most of the land in question had been zoned for agricultural land uses or for very large lots for residential uses. The developers decided that the highest and best use of the land would be for apartments and townhouses of the rental and condominium varieties. Some of the plans called for innovative developments combining apartments and townhouses with single family residences and shopping areas, albeit at high densities of land usage. The zoning ordinance of the county in which the central city and most of the land in question was located had to be revised in order to allow such developments since most of the area was zoned for farming.

At first few people, including the planners, realized what was happening. Some of the zoning changes were relatively small; others were substantial but of the planned community type of development which was popular with planners. As it became clear that there was indeed a movement to develop as much of the river area as possible, at high densities and regardless of the loss of natural amenities, reactions began to occur. The planners proposed to formulate a development plan for the area along the river based largely upon a definition of the river basin as that of its 100-year flood plain. At the same time a developer filed for rezoning to allow him to develop a planned community along the banks of the river which would have one of the largest single developments in the history of the south.

Sensitive to the urgency of the situation, and prompted by the commissioners of the county commission, the county and regional planning department staffs began a crash program to formulate the river basin development plan. They were convinced that a sound technical plan that was reasonably scientific and rational would serve as the guiding document for the decisions on rezoning and provision of water and sewer facilities being requested by the large developers. The project staff worked overtime and weekends to collect the appropriate information to allow the problems to be delineated and interrelated. Following the traditional planning process, they formulated several alternative development schemes and analyzed each. A conclusion was reached as to the most appropriate development pattern to be recommended, and this was presented to the county commissioners and the community.

The plan was hailed by the media and various community leaders. The county commissioners remarked that they were studying the plan and were impressed with the obvious amount of time and energy that the planners had invested in the effort. Several neighboring counties stated that they would accept the plan in principle as a basis for decision-making, and most of the environmental and neighborhood associations interested in the problems went on record in support of the plan. Through the somewhat elusive process of trying to define the community objectives for the area, as well as attempting to describe the problems and potentials, the planners had devised a plan that was acceptable to several special interests and many of the opinion leaders from the press and mass communications media. The application for the massive rezoning, however, was still pending.

The rezoning application for the gigantic development along the river was in violation of the basic intent of the plan to limit development by relatively low density standards so as not to disturb the ecological balances and the recreational and scenic values associated with the river. This would result also in public savings for water and sewer facility extensions. The rezoning application came before the planning commission which sat as the independent policy board over the professional staff. As was customary, the citizen planners on the planning commission submitted the application to the professional staff for review.

As has been the historical role of the independant planning commission, it attempted to shield the planners from the political implications of the river case. A secondary role for the independant commission has been to represent the interests of the community, however ill-defined these interests may be. Comparing the rezoning request with the technical aspects of the plan, which had been accepted by the county commissioners as a reasonable guideline for decision-making, the professional planners recommended that the rezoning be denied. They cited the high density, large number of dwelling units, withdrawal of recreation land now used by the public (although there is and will continue to be debate over who actually owns the river, the state or the abutting property owners), and some vague criticisms couched in terms of too much development and too much environmental damage. The recommendation was rational in terms of the planning principles that had gone into the preparation of the plan. Under considerable pressure from the special interest groups of environmentalists and affected neighborhoods, the planning commissioners concurred with the recommendations of the professional staff. The only opposition had come from some developers and members of other economic groups who did not wish to enter the debate in the forum of the public hearings held by the planning commission. The recommendation for rejection of the rezoning proposal was sent to the county commissioners for final approval.

The planners, both professional and citizen commissioners, had acted within the context of technical merits which were based upon general principles of planning practice. The only special interest groups with major support for the rezoning were involved in the planning process, albeit low key, through the public hearings to defend the project on its economic benefits. The recommendation now was thrust into the political arena where elected officials had to make the final decision based upon the balance of supports and demands.

The county commissioners had the professional staff and planning commission recommendation in hand as they received the input from the special interest groups. The environmental groups had been unpopular with the county commissioners because they tended to be dominated by white, middle-class intellectuals with little political or economic clout, at least clout that could be applied for reasons of decision-making in government. The neighborhood groups affected were somewhat divided with some owners seeing destruction of their homes and property while others saw their opportunity to make a fortune by selling land bought years ago for a pittance or inherited from land-poor ancestors. The developers formed a coalition since, if one were victorious, a door would be opened. Many of the economic influentials, such as realtors, bankers, and investors, tended to coalesce into an effective special interest group around this issue as they had on other issues. Ironically many of the leaders of the environmentalist groups were the wives of the leaders of the economic influentials group. Amongst all of this the low-income and black organizations of the central city saw the entire situation as white hypocrisy because they thought only well-to-do suburbanites were interested in preserving the natural state of the river. These latter special interest groups tended to imply a latent support for the rezoning since it would mean at least some jobs.

The planning commission and its staff were unaware of the amount of interest in the river case. Fanned by smoldering editorials against the rezoning by the local papers, the planners could do little but fall back on the technical matters of density, utility services, and pollution dangers. The citizen commissioners preferred that the professional planners handle this release of information and abdicated any further role in this matter. The professional planners were emerging incredibly as local folk heroes in the battle to save the river for despoilation.

There were three county commissioners, one of whom acted as chairman and usually voted only in the event of a tie. When the matter came up during a county commissioners' session packed with all interested parties, it was deferred; in fact, it was deferred for three months supposedly on the grounds of the need for better and clearer information from the planners. A vote was finally called and resulted in a 1 to 1 tie. The chairman of the county commission cast his vote for the rezoning, and it was approved. A storm of protest went up in the press and in the civic association meetings; the developer started his earthworks as well.

The elected decision-makers were criticized as being irrational and mindful only of special interests. The first part of this criticism is nonsense since rationality is a variable function which is affected by the constraints and parameters of the system. In the political system that has been described at the local level in previous discussion, their actions are rational more or less. The three commissioners balanced the special interests supports and demands along with the planners' input and reached conclusions. This is rational within the structure of the political process. The special interest groups opposed to the rezoning did not have sufficient equity in the minds of two commissioners, and the planners opposition was solely in terms of principles and theories which held apparently no equity position with any of the commissioners. The special interests in favor of the rezoning were economic influentials, campaign contributors, opinion leaders, and builders who brought jobs and new money to the region. The complex question in the minds of all three commissioners was whether or not the press and environmental groups had raised a sufficient fuss to affect future voting patterns of the general public. Two of the three commissioners obviously felt the future for their political careers was not periled. The threat of intervention from the state or federal level was serious, but the three commissioners felt this to be unlikely.

The planners could have done more in this case. They were obviously somewhat naive and unaware of the political process and probably believed that the soundness of the technical merits exuded by the planning process were sufficient and necessary for victory. The planners assumed that the citizen members of the planning commission represented the special interests of the region and it was not necessary to pursue the equity positions any further. Furthermore, the planners seemed opposed to altering their technical recommendations and were content to lead an all-or-nothing decision-making battle. Most astounding of all were the public utterances of the planners to the effect that they should not be involved in matters other than technical merits and that as technicians they should not consider compromises in what they had determined to be the best plan. When planners are ineffective they often tend to substitute arrogance, and they scorn the machinations of politics and special interests. While such a substitution is stimulating intellectually, and is sufficient for conferences and cocktail parties, it does not efface the ineffectiveness. The river case is but one illustration of how the planning process and the political process seem to be out of fit.

Techniques Used in Planning

Since there is a reason to believe that the planning process is difficult to relate adequately to the political process in local government, it is interesting to examine the techniques that planners use to arrive at conclusions to serve as a basis for recommendations. The politician's techniques are relatively intuitive, based upon his experience, judgment, and skill at assessing the mood and attitude of the electorate. The planners have been attempting for many years to make their techniques less intuitive and to base them in scientific and quantitative factors to a great extent.

Since the profession of planning has roots in the architecture and related design fields, it is not surprising that the early techniques used by planners were largely artistic and intuitive. The early techniques used somewhat utopian concepts to formulate the basis for the master plan which was to serve as the overall guide to physical development in the community. Planners were quick to point out that the plan was a guide and not a blueprint, and the plan was relatively flexible when it came to specific projects. As long as the overall thrust of the master plan was retained, there were allowable variations.

Most of the significant planning activities through the first half of this century were concerned with the development of master plans. The federal government latently provided grant programs for this purpose under the Housing Act of 1954, Section 701. For a short period this new transfusion of federal dollars led to a sizable interest in master plans, but the peak of the utility of this technique was past. The master plan technique, coupled with its usual implementing tools of zoning, subdivision controls, and long-range capital improvements program, had produced incredibly few successes. The absence of substantial citizen and group participation, the elitist notion of an implied perfect plan which all must accept, and the paucity of quantitative tools used in formulation of the master plan were major criticisms.

Toward the end of the 1950s and during the early 1960s, two major changes occurred in federal programs which impacted planning techniques at the local level. The first change was the broadening of the concept of a single master plan into that of several functional plans which could be integrated into a comprehensive plan. These functional plans were for such local activities as roads, parks, health centers, schools, and utilities which could then be interrelated and coordinated by planners using the comprehensive plan technique. This technique was useful for bringing the several functional plans into one source of policy and guidance; the planner was given the role of comprehensive synthesizer, which was a variation, if not a complete change, from his traditional role. The major difference was that, while the planner did not necessarily develop all of the plans for local government, he did attempt to bring them together. The techniques of the planners began to enter the area of policy analysis and review which necessitated the inclusion of more scientific techniques.

The second major change of the late 1950s and early 1960s was the emergence of a demand, if somewhat artificially induced by the federal government, for direct participation by citizens (more so than groups) in the planning process. Most of the planning sponsored by federal funds insisted upon elaborate mechanisms for involvement of as many citizens as possible in the planning process, even to the extent of asking citizens for advice and assistance on relatively technical matters. Because the tradional techniques of planners made small allowance for large-scale involvement of ordinary citizens, the immediate effects were chaotic and counterproductive. Some semblance of reason and constraint was garnered, and guidelines were changed in the late 1960s so that the citizen participation requirements were made more flexible and less binding. Nonetheless the planners had been thrust into areas of public exposure that required new techniques. Equally interesting was the result that citizens were beginning to look more toward the planners for technical and scientific information.

The recent trends in techniques for planning at the local level have been extensions of these events. The techniques of planning are becoming inclusive of communications and dissemination of information to citizens and groups interested in planning problems. This necessitates a review of political and social variables by the planner. The other technique set is inclusive of more quantification and a quasi-scientific orientation. The latter techniques are more popular with planners than the former.

The call for more scientific content in local planning has emanated largely from the success of the so-called systems approach by engineers and scientists most spectacularly in the aerospace program. Politicians, teachers, businessmen, and students, as well as the various special interest groups, have formulated a question that may well be the platitude of modern times: "If this country can land a man on the moon, why can't it solve city problems here on earth?" Because the systems approach was so efficacious in landing an astronaut on the moon, why not use this set of techniques for solving the problems of local government: only the locale is different, argue many people. Planning techniques must become primarily scientific, it is implied, and this will provide the basis for better implementation.

This country can land a man on the moon and probably cannot solve its city problems, at least if such a hope is based primarily on the inculcation of systems techniques. The irony of this movement is that it is conceptually the same as the master plan techniques except that scientific elitism is to be used in place of intuitive elitism, and there is some disagreement over whether the two forms of elitism are different at all.[7] The movement to make whole cloth transfers of systems techniques and related scientific and engineering approaches is incredibly popular but astonishingly dismal in results. The basic reason is that the basic premise is erroneous. Changing the context of application of a set of techniques has remarkable implications for the usefulness of techniques. The development of new disciplines of techniques without substance is intellectually dubious, but far more serious for local planning is the threat that many of the right answers are being set forth for the wrong questions.[8]

The political parameters of the movement toward new scientific techniques for planning are interesting. It is clear that planners must become more politicized and offer more technical content to their proposals, but the sheer brute force transfers of systems techniques into planning are doomed for several reasons. An interesting way to examine the reasons is to compare the dichotomies of the Apollo Mission and urban mission as listed in Table 1.

Table 1
THE DICHOTOMY OF THE MISSIONS

Apollo Mission	Urban Mission
Clear goals and objectives	Unclear goals and objectives
Chain of command	Splintered decision-making
Authority and responsibility	Authority with little responsibility
Adequate resources	Inadequate resources
Technical problems	Nontechnical problems

The Apollo Mission had a very clear goal. For example a basic goal was enunciated by President John F. Kennedy in 1960: "This country will land a man on the moon by 1970." Such a clear-cut goal could be translated readily into action-oriented objectives for accomplishment. There was little citizen participation in setting this goal, nor were there elaborate efforts to convince the electorate that this was a wise use of resources. This was a goal that was related to the honor and primacy of the country and had to be attained—literally at any cost.

The cities of this country have never fared so well. There are no clear goals for state and local government set from on high. The goals that have been used for planning purposes are essentially statements of support for the principles of American motherhood, hot dogs, and apple pie. When the local planning agency sets forth the goals of a healthy, pleasant, and attractive city, few people will argue. Nor is it debatable that the community should be free from crime, disease, and ignorance. Such goals are meaningless in terms of setting forth a basis for formulating objectives, however. The only utility of unclear goals and objectives for planning is that they are sufficiently vague and so noncontroversial that the planners need not be impeded in their activities by having a specific direction and exact purpose for plans.

The hierarchy for decision-making in the Apollo Mission was relatively straightforward. A chain of command that was a combination of military order and business efficiency was structured so that the brunt of decision-making was clear to planners and scientists working on the project. There was little doubt as to where a technician would turn for approval of a recommendation. When mistakes were detected, the person accountable was apparent; rewards were given correctly when achievements were made in decision-making as well.

Decision-making in local government is at its best confused at its worst nonexistent. The chief executive may not always be in charge of affairs. Differing and changing strengths of special interest groups cause variation in who influences a given decision. Tensions between legislative and executive leaders often lead to clouding decisions, and recent patterns of court decisions bordering more upon unrestrained judicial initiative have created an additional governmental tier of decisions. That there exists any semblance of a well-structured chain-of-command for decision-making at the state and local level is dubious. The planner is faced with the remarkable task of determining to whom he should propose his plans; a wrong choice could be dramatically suicidal.

The lead organizations in the Apollo Mission, such as the National Aeronautics and Space Administration, were in the enviable position of having both the authority and the responsibility to reach their goals and objectives. This meant that NASA had been delegated the authority to undertake mission planning and programming, as well as the responsibility of implementing the plans and programs. Authority and responsibility were integrated in a single organization which led to many significant examples of speed and efficiency in plan fulfillment.

The planners in state and local government have never had the responsibility of plan implementation, nor are they likely to ever assume such a power. Planning in local government is still an undertaking in which one or more organizations are authorized to plan, while differing organizations are responsible to effectuate plans. Most state and local governments are unwilling to combine plan implementation responsibilities with planning authorizations for fear of creating superpowerful centralized agencies, a fear that has been traditional in the United States.

It could be argued effectively that the resources from the public treasury made available for the completion of the Apollo Mission were adequate. In addition to the direct resource allocations, there was further executive ordering of resources to be directed from related military and general governmental operations. Contractors that were bidding at apparently inflated prices were not reprimanded, because the success of the mission was more important than some excessive profit margin considerations. The facilities sponsored by the federal government were ornate by most standards, and the spinoffs of Apollo Mission research and development for universities, research institutes, the military, and industry were sufficient to generate more than desirable interests and supports.

Resources for the implementation of plans for state and local government were never adequate, nor are they likely to be adequate in the future. Much of this can be related to a predominance of rhetoric with omissions of commitment; that is to say that many decisions have been made for solving urban problems by state and local governments, but there has rarely been the corresponding action program commitments. Planners and members of related professions have not been especially realistic in their estimates of the resource needs, and many estimates are so high as to almost guarantee that reasonable allocations will not be made since the costs are unthinkable.[9] Basic to many of these problems is the absence of an acceptable concept of which costs are public and which are private. Many costs have been implied as pure public (e.g., highways), while some are considered private (e.g., airlines), yet there is little conceptual basis for the melange of resource commitments pertaining to most other functions. This undoubtedly has to do, to a great extent, with the ambivalent mixture of public and private capital required for plan implementation, as well as with the changing structure of capitalism in America. Thus in addition to determining how much is needed, the planner must determine where and how to get it in the absence of any acceptable process of fair share payments.

The major problems that NASA planners had to face were concerned with development of technological improvements to enable spacecraft and crew to be launched, guided, sustained, and returned to earth. By and large these were technical matters having to do with thrust, torque, fuel, life-support systems, and hardware. There were some problems of psychological conditioning and training of astronauts, but even these were relatively technical concerns.

The problems of state and local government pertinent to planning are rarely technical. On the contrary it appears that the technology available to local governments, be it computers, automation, construction, materials, or transportation equipment, is at a level far beyond the capabilities of local governments to utilize it. Rapid transit systems can be built; prefabricated and industrialized housing can be built; information systems can be developed; communications systems can be deployed; and utilities and services can be automated according to the highest technological specifications. This is not the gist of urban problems at the state and local level. The gist of these problems is social, political, and economic—people problems that defy automation and standardization.

These dichotomies between the Apollo Mission and the urban mission show that the context of each was different to a substantial degree. The Apollo Mission had all of the prerequisites for effective and efficient decision-making and commitment while the urban mission has few. It would be intriguing to argue that these conditions could be changed so that systems techniques could be applied to planning at the state and local level as several supporters have argued, but it seems incredibly naive to think that Americans will restructure radically their government in order to allow better techniques of planning to be instituted. It seems even puerile to consider these matters as crises which seriously disturb most Americans who are content to allow government to muddle through its affairs.[10] While there are efforts under way to change the adverse situation of decision-making at the state and local level, it seems reasonable to assume that techniques suitable for the more orderly and structured context of the Apollo Mission are of limited value for the urban mission unless modified substantially to deal with the different set of conditions.

The influx of scientific methods into the techniques of planning can only be valuable if there are modifications and recasting of these methods to handle the idiosyncracies of local government. This will mean that the optimal basis for scientific solutions to problems will have to be inclusive of qualitative as well as quantitative factors. Whole cloth transfers from systems analysis will never be successful in local government unless elements of the political process and its equity positions are introduced into the calculations. The planner must emerge as a scientist who understands and recognizes the political process and can utilize scientific tools in such a manner as to strengthen the basis for decision-making. Further attempts at ignoring the political process and attempting to replace it with scientific techniques are bound to result in considerable waste of resources and efforts. There will undoubtedly be more scientific techniques used in planning, but these will be hybrid techniques borrowed from other fields and modified to fit the political process and planning practices.[11]

The PPBS Case

An interesting case study of the influx of systems techniques into planning is found in a state planning agency in a western state. The agency had been making considerable progress toward formulating plans within an appropriate political context which was leading to achievements uncommon in other states; the state planning agency was identified generally as one of the most effective in the country. The state was beset with unemployed aerospace engineers and a military-industrial complex that had been molting as a result of cutbacks in the defense and space programs. The professional and corporate special interests exerted particular pressures upon the governor and legislature to utilize their system approach techniques in state planning.

A few opportunists in state government saw this as an opportunity to make their fortune. They joined forces with several of these special interests and devised a planning-programming-budget system (PPBS) which incorporated techniques of aerospace and defense industry planning. They used the familiar argument that it had been successful in these other fields of public interest, and it would obviously be effective in state planning. The so-called PPBS was cast in the rather extraordinary form of a bill for submission to the legislature. The reason for this tactic was the insistence on the part of the state planning staff that the PPBS be modified to require less mathematical data and incorporate more qualitative and experiential factors. This position was unacceptable to the framers of the bill because they saw a potential for rising above the state planning staff and becoming the central planners for the entire state.

The state planning staff had a difficult strategy selection problem. They could have lobbied against the maverick bill which would have the effect of making system approaches to planning a legal requirement. This would have meant a confrontation with powerful special interests as well as the danger of creating an image of being old-fashioned bureaucrats uninterested in new ways. An alternative strategy would have been to coopt the movement by instituting a PPBS voluntarily and thus become involved in using system analysis techniques prior to any requirements. The third and eventually employed alternative strategy was to do nothing and wait for the systems approach advocates to create their own quagmire and go down in it. This is precisely what happened.

While key legislative leaders and the governor knew of the reluctance of the state planning agency to incorporate this whole cloth transfer of systems analysis from defense and aerospace industries into state planning, the absence of any articulated and vociferous opposition resulted in the act's being passed. Incredibly the act went into details of data specification and types of cost and benefit models to be used for analysis. An organizational structure that essentially placed the PPBS staff at the center of state government was included. An edict was included that required complete cooperation of all state agencies and local planning groups. The bill was more of a specifications program for a rocket launching than the formulation of a planning process, but it was passed.

Within the astonishing time of a few years, the PPBS had collapsed. The requirements of the law could not be met because data were unavailable and technical problems of quantification and parameter estimation of the analysis models could not be completed. Even though required under state law to be fully cooperative, none of the state agencies would participate in earnest in a project they viewed as a threat to their power base and an effort by a central elite group to take over all decision-making and resource allocation. The systems analysts and consultants employed in the program began to leave as frustrations mounted and confidence evaporated. The state planning agency gradually reassumed its lead in policy analysis and began to employ on a moderate and temperate level many of the techniques that could be modified from systems analysis. The state planning staff was successful in this battle because the whole cloth transfer of scientific techniques into government planning killed itself.

The state planning agency in this case is now on the way to making a potential contribution to the field. They have weathered their most serious challenge and have retained the confidence of political leaders. They are convinced that there is some value to forming hybrid models of analysis borrowed from other disciplines if these models can be recast in a political process context with social and economic variables. They realize that scientific techniques can be modified to be used as tools for improving planning but not used as surrogates for political decision-making. The state planning agency has devised an approach for living with the influx of quantitative techniques into planning while maintaining effectiveness in the political process.

NOTES

1. For a thorough review of cybernetics, see Norbert Weiner, *The Human Use of Human Beings: Cybernetics and Society* (Boston: Houghton-Mifflin, 1954).

2. A detailed theoretical and applied science treatment of this subject is available in A. J. Catanese and Alan W. Steiss, *Systemic Planning: Theory and Application* (Lexington, Mass.: D. C. Heath, 1970).

3. The landmark work in this area is David A. Easton, *A Systems Analysis of Political Life* (New York: John Wiley, 1965).

4. A good analysis of the evolution and problems of Advocacy Planning is found in Earl M. Blecher, *Advocacy Planning For Urban Development* (New York: Praeger, 1971). For an ultimate view of radicalized Advocacy Planning as a steppingstone to "community socialism," see Robert Goodman, *After the Planners* (New York: Simon and Shuster, 1971).

5. This matter was discussed thoroughly in the classic work Martin Meyerson and Edward C. Banfield, *Politics, Planning, and the Public*

Interest: The Case of Public Housing in Chicago (New York: The Free Press, 1955), Chap. 9.

6. While he has made this statement in various works, the basic reference is Coleman Woodbury, ed., *The Future of Cities and Urban Redevelopment* (Chicago: University of Chicago Press, 1953).

7. Many writers have examined the serious questions of values and elitism in science, and one of the best such examinations is C. West Churchman, *Prediction and Optimal Decision: Philosophical Issues of a Science of Values* (Englewood Cliffs, N.J.: Prentice-Hall, 1964).

8. Many of these points are extensions of earlier work with Alan W. Steiss published as Catanese and Steiss, op. cit. Recent work has been a textbook on scientific methods used in planning in the manner discussed. See A. J. Catanese, *Scientific Methods of Urban Analysis* (Urbana: University of Illinois Press, 1972).

9. Several examples are found in the literature of the field. With regard to the urban renewal program, it has been argued that a substantial share of the Gross National Product would have been required for actual implementation. See Martin Anderson, *The Federal Bulldozer* (Cambridge: Massachusetts Institute of Technology Press, 1964), Chaps. 8 and 10. Similarly it has been argued that many improvements proposed by planners are acceptable to citizens only when there are adequate federal or state funds available but not if they require direct payments or taxes raised from local citizens. See Edward C. Banfield, *The Unheavenly City* (Boston: Little, Brown, 1970), Chap. 1.

10. It has been propositioned that public managers and planners can nonetheless work within this dilemma, because they exude an aura of scientific method and objectivity but are really like the public-at-large and seek to raise crises only when they are in their own self-interests. The classic work is Charles E. Lindbloom, "The Science of Muddling Through," Public Administration Review, Vol. 19 (Spring, 1959), pp. 79-88.

11. An interesting analysis of this trend is found in Britton Harris, "Foreword," in *Decision-Making in Urban Planning: An Introduction to New Methodologies,* ed. Ira M. Robinson (Beverly Hills, Calif.: Sage, 1972), pp. 9-20.

There is a story going around about the Almighty assigning a team of Angels to collect the parts so that he might make a new Adam and Eve and attempt to improve upon his past failures. The Angels enter the central warehouse and ask for assistance in purchasing the various parts. The Chief Supply Angel directs them to various locations and accounts for the sums spent, but the Angels reach an impasse when it comes to brains.

The Angels find that there are several types of brains available at differing costs. They notice that a pound of architect brains costs $10.50; an engineer's brain costs $15.00 per pound; a planner's brain costs $19.95 per pound; and a politician's brain costs $199.99 per pound. Somewhat aghast they summon the Chief Supply Angel and ask why the politician-type brain is so expensive.

"Do you realize how many politicians it takes to produce a pound of brains?" he replies.

This story is not too far-fetched a comparison of the way that many planners feel about state and local politicians. They often find that many other professional people and writers share their opinion. The politician is regarded commonly as the imbecilic finagler who has wormed his way into office for the worst of all reasons. The fact that he was elected is only evidence of the sad state of public wisdom.

Such a prejudice is common among many middle-class intellectuals. The politician is assumed to be somewhat incompetent by virtue of his choosing such a career, and the best that can be done to resolve such a hopeless situation is to develop a core of middle-class professional bureaucrats who can advise him on what to do for the community. A related bias has been to develop a movement toward stronger decision-making by the chief executive because he has access to the professional bureaucracy and can be held accountable more directly to the polity. This doctrine has for many years been a significant part of the dogma of public administration and its various specializations, although there appears to have been a split among many theorists for the last couple of decades.[1] One of the more unfortunate results of this prejudice has been a fairly entrenched notion among planners that politicians are incapable of making a correct decision on their own. This is not a useful basis for improving the effectiveness of planning.

Politicians tend to be suspicious by nature, and it is difficult to conceal contempt from them. Many politicians are aware that planners, as well as architects, engineers, and other professions working with planners, are prejudiced against them and adopt a defensive posture. This posture often means an automatic cataloging of recommendations from planners as being impractical, unrealistic, and irrelevant. From there a routine begins whereby the planner attempts to convince the politician that his input to the political process is none of these, and he really has the same interests at heart as the politician. When the planner is unable to get his recommendations accepted, he often bemoans the "lack of understanding" that exists among politicians.[2]

Submerged among these problems of communication and understanding is a set of basic dilemmas. These revolve around the profound differences between the personalities, training, experiences, and ethics of planners and politicians. The situation is complicated further by hassles among planners and other professions and nonprofessions as to who are the real planners. The apparent result of this set of dilemmas and controversies is a movement toward new kinds of characteristics for planners in terms of training and ethics, as well as efforts to help the planner understand the politician.

Aspects of Personalities

Examination of such a complex feature of politicians and planners as personality is difficult and of necessity requires generalization. The distinctions and exceptions found among both groups are obvious to even the most casual observer of state and local government. Nonetheless it is essential that generalized discussion of the key points of difference between the personalities of planners and those of politicians be undertaken if improvements in the planning and political process are sought.

At a recent national meeting at which a mayor from a large city and a leader of the planning profession were on a panel, a student from the audience asked a most curious question: "If you had not gone into your present field (mayor or planner), what would you have become?" The mayor responded quickly that he would have specialized in corporate or tax law with a large law firm or would have been an executive in a large business. The planner hesitated pensively and then replied that he would have been a leader in the Boy Scouts of America or a missionary to an undeveloped country for his church. The student seemed satisfied with these replies and launched into a castigation of the planning profession for being composed largely of Boy Scouts and missionaries who were ineffective in dealing with the hard-bitten lawyers and businessmen who run government. The scene was almost too perfect and has been a source of much debate in the last few years. Many planners have argued that such an alternate career choice is not representative of younger planners and many of the veterans of the planning profession. Yet the unrehearsed response of the planner to this interesting question displayed elements of personality traits that may not be so unusual among planners.

The politician tends to be attuned to Calvin Coolidge's dogma that "the business of America is business." He comes out of the business community or the legal profession in general. There are significant examples of farmers, physicians, and teachers in politics, as well as a growing number of housewives, but businessmen and lawyers still predominate, especially in the legislative branch. This does not necessarily mean a conservative, elitist power structure as the early community power studies proposed, but it does tend to signify a moderate political philosophy and commitment to a Judeo-Christian Work Ethic and self-improvement.[3]

Certain personality traits are common for the politician. He tends to be somewhat provincial in the sense that he is bound by his roots to the state and local area in which most of his career will take place. As long as he is a state or local politician, he may put aside opportunities for changes in his life style that may involve geographical change or increased mobility. There are some significant exceptions in states with high growth rates, such as California, Florida, Hawaii, and Arizona, but the general rule is that the politician is born, raised, and buried in the same area. He is somewhat anachronistic in this sense because many Americans have adapted to high mobility as a way of life. Perhaps because the personality of the politician can resist the temptation to pull up roots and replant them in a better place, he offers a stability and permanence that is vital to the structure of state and local government. In rapidly growing suburbs of major cities, a candidate who can proclaim that he "was born and raised in this county before there were any people here" conveys an image of steadfastness that appeals to the new residents. There may be some historical basis for this phenomena as in the sad case of the "carpetbaggers" of the post-Civil War era in the south. All things being equal, the newcomer will generally lose a political race to a native largely because the former's motives for moving are suspect.

Retention of political power is of paramount importance to the state and local politician because of this personality trait. "Getting reelected is the first rule of politics" is ingrained in political life. It may even supercede immediate opportunities for advancing in the political structure if there is danger that the advancement may result in loss of existing political power. This condition directly affects the kinds of decisions that the politician may make. It often means that friends and supporters have a special equity in the political process. This could mean that the politician can be an idealist or adhere to a specific political viewpoint to only a limited degree. He must adopt a posture of being a "satisficer" so that the major equity positions are content and the minority equity positions are not disgruntled entirely.[4] If this entails the "art of compromise," then it is rational to the politician to behave in such a manner. The personality traits of the politician are such that he can develop a corresponding set of ethics to accommodate the needs for retention of power and stability. It is the rare politician at the state and local level who feels bad over a compromise decision. On the contrary an ability to construct compromise decisions is considered a remarkable trait of the politician.

Planners at the state and local level have aspects of personality which are different from those of the politician. A substantial factor of this set of differences may be linked to personality trait of the planner which compels the need for change and mobility. The planner thrives on changes in communities and wants to be a key element in affecting the nature of that change. He welcomes mobility and tends to move often from position to position in different states and cities; indeed, it is the rare planner who remains in one place for several years. The mobility of planners and related professionals is so great that a theory surrounding the concept that territorial space is somewhat limiting as a basis for planning and a nonplace urban realm is more useful has been derived and embedded in planning theory.[5] While the profession of planning is localized in the governmental hierarchy, the professional planner is highly mobilized in geographic as well as social and economic terms.

State and local planners are no more or less ambitious than any other group. They want to attain success and recognition and all that comes with achievement. Planners differ from politicians, however, in that they usually will describe their careers as being dedicated to "helping the community become a better place" or "leaving society a little better off when I am gone." The planner's efforts are group-oriented rather than self-oriented. Some believe that this is the result of his having been raised in middle-class America without ever having to struggle to survive and thus being less inhibited by the need for status and wealth. This may be a trait that is common for many planners but not for all since recent years have found an influx of status-conscious minority group members into the profession. Yet there is sketchy evidence that even the minority newcomers have come from comparatively middle-class segments of the Black, Chicano, and Puerto Rican communities and can avoid somewhat the need for personal wealth and status.

The traits of mobility and group-orientation derived from the personalities of many planners have cast upon them the stereotype of humanitarians and idealists with part measure of scientist. The stereotype is correct to only a limited degree. The group orientation of planners forces them by virtue of their personalities and by the realities of sponsorship to seek client groups. The major schism in the planning profession and among its critics and friends has been whether planners have sought clients in an elitist or populist way. Some argue that planners have selected their clients in such a manner as to ensure the follow-through of their middle-class and intellectual roots which sometimes would include politicos.[6] Others have argued that the planners can only succeed by offering their assistance to groups that cannot elsewhere obtain planning advice.[7] The actual conditions are such that both traits are manifested in the real world and this is in part why plans are cluttered: they reflect pluralist objectives of the community. Thus the so-called humanitarian and idealist traits of the planners exist in relation to the client selection process practiced by different types of planners.

Do planners seek political power? Probably not in the normal way and certainly not in the sense that politicians seek to acquire and retain political power. The personality of the planner is such that he can accept the loss of glory and fame that comes with political successes and substitute the satisfaction of knowing that one's ideas and concepts have been employed in bettering the community. In a sense this is a kind of political power, albeit covert and tentative, and it appears to be sufficient for most planners. It is extremely doubtful that most planners would welcome efforts to make planners more responsible directly to the polity through election or any similar move to bring about elements of held political power by planners largely due to this personality trait (even though professionalism usually will be the reason cited).

A similarity between the personalities of politicians and those of planners is their inherent suspicious nature. The politician, as mentioned, is suspicious of intellectuals, liberals, and public interest groups. The planner is suspicious of businessmen and investors. This trait may be one that is instilled through education and experience with other planners, or it may be a mild rebellion against his upbringing by middle-class parents who were probably in business. Throughout the literature of planning, the canons of the professional society, and the recorded efforts of practitioners, there exists a common thread of striving to fulfill the public good and not allowing private greed and avarice to prevent same. While many intellectuals have called for big business to devote a portion of its resources to solving urban ills, nowhere has the call been viewed more suspiciously than among state and local planners. Curiously like their early predecessors, however, planners may criticize capitalism and big business but do not seek to foster different or new economic systems in America. Thus the trait is more a personality quirk than a doctrinaire stance.

The Sewer Case

An interesting case study of a clash that arose largely because of personality differences between planners and politicians concerns the approval of a water and sewer system in a Midatlantic state. Under requirements established by the federal government, the state planning office had to approve plans for the construction of water and sewer systems made by local planners. This was to ensure that population settlement patterns, demands for services, and economic development efforts were coordinated by the local government planners and politicians. There were no special guidelines as to how the analysis should be undertaken, and most states formulated their own approaches.

The particular state planning agency under discussion had not formulated any real set of procedures, relying instead on the collective experience and wisdom of the professional planners to establish a basis for approval or disapproval of local plans. The state planners believed that orderly patterns of growth could be established and articulated to a satisfactory degree to serve as criteria for analysis of water and sewer systems.

A small town from an impoverished part of the state developed, with the aid of consulting planners and engineers, a plan for a water and sewer system. The plan was ambitious in the sense that it provided for more capacity and service than could be immediately used and was based on the premise that a relatively aggressive industrial development effort by the town would be enhanced by the new facilities. Water and sewer services were extended under the plan to a relatively vacant area of the town which was designated as the industrial center on the town plan and zoning ordinance. The water and sewer facilities would be paid for by a combination of federal grants and bonds floated by the town. The consulting planners believed this a reasonable set of objectives and so devised the general plan which was in turn given technical specifications by an engineering consulting firm.

The application was denied approval by the state planners because they believed that it was poorly designed and was not in the best interests of the community. They argued that the residents would have to pay debt service on the bonds for water and sewer facilities that would most likely go unused for several years. Furthermore they believed that the growth pattern that would follow the water and sewer facilities was not in keeping with principles of sound city planning because it would result in what they labeled as "sprawl." For reasons that are unclear, the planners did not outline any alternative plans nor did they give any indication of what had to be revised in order for the application to gain their approval.

The mayor of the town and several of the councilmen and businessmen knew the governor of the state. Within two hours of receiving the denial notice, they had arranged a meeting with the governor. The town had voted overwhelmingly for the governor, and he considered it much like his own hometown. These people were his friends and supporters: they were important to him. The governor in general was a great supporter of planning and had reorganized the planning agency to place it in his office close to him personally. He called his top state planners to his office to explain the problem. After listening to the reasons stated on the record and further expansion, the governor said that he thought the planners were being somewhat narrow in their interpretation of the federal review and approval program. The planners argued that the federal officials were their clients in this matter since they were in essence delegating their authority to the state planners.

The delegation from the town met with the governor and explained to him their support and admiration for him personally and his programs. They felt, however, that the denial of their water and sewer system application was contrary to his stated objectives of economic development and revival of the declining small towns. The mayor said that the planners were just being "intellectual dilettantes" and had no real basis for denial of the application other than some vague notions about principles of city planning. The mayor stated that the entire town would be most displeased if the water and sewer program were canceled since it would curtail their program of attracting new jobs to replace the declining unskilled jobs in the town.

The governor asked the state planners to reconsider the application. Sensing this as political pressure to approve the application, the state planners decided at a staff meeting to deny again the application but offered to assist the town in revising its plan. They insisted that the director of the planning agency convey this message to the governor and let it be known that they would not succumb to political pressure or what they termed "chronyism." Professional planners, they inveighed, were beyond political considerations and had to act in a highly professional manner. Federal officials were attempting to increase the effectiveness of state planning by delegating this approval power to the professional planners, they wrote, and it was up to them to merit the confidence of the federal government.

The governor blew his stack. He told the director of the state planning agency that he was their one and only client, and that meant his friends and supporters in the small towns of the state. He said that these were poor but hardworking people who were trying to improve their town and here his own advisers were trying to sandbag the efforts. The governor reminisced about his days as a small-town mayor and the incredible problems he had with federal and state officials. This was not going to happen in his administration. If the planners could not serve the people, then there would not be any planning. He concluded his tirade by saying that he would abolish the agency by executive order and fire the director and all professional planners if the water and sewer grant application was not approved. Two weeks later the application was approved.

Two months later the director of the state planning agency was asked to submit his resignation and the agency was reorganized. The new director was a young businessman who agreed with the governor that planning had to be related primarily to the current administration. This meant that political considerations related to supporters of the administration were important variables in planning. It also meant that planning should be action-oriented and based on real life problems rather than textbooks on city planning principles. This did not mean that planning had to be unimaginative and unexciting; on the contrary, the young businessman stated, the governor had exciting and ambitious ideas for the state which could be achieved through state planning.

The governor liked the way this fellow talked. The state planning agency has made substantial gains in the last few years since the reorganization and personnel changes. Most of the persons hired as planners have had no experience or training in the field of city and regional planning but have come from law, business, and public administration backgrounds. Apparently this was the wish of the governor and several of his top aides. The new director concurred with this suggestion.

The Reapportionment Case

A major city and its suburbs in the south were ordered to reapportion all of its political districts to reflect changes in population shown by the census. The federal court suggested that the city planning department had the available information and skills to undertake this chore. A joint committee was structured by the state legislature and city council to oversee this effort of the planning department. The political parties created reapportionment committees of their own even though it was a one-party area. Several public interest, minority, and voter education groups expressed an interest in the reapportionment proceedings as well. The chairman of the joint legislative committee met with the planning director, who was reluctant to accept the problem to begin with, and suggested that reapportionment was the most critical issue among politicians since it involved retention of office. He suggested that the most acceptable plan would be one which preserved the districts of most officeholders who expressed interest in returning to office but changed their boundaries in such a way as to ensure proper population figures. It was also suggested that the central city blacks had ample political equity in the reapportionment question and should be consulted. The various public interest groups should be heard but not taken too seriously. The chairman said he was confident that the planners could use their data and computers to draw the new maps.

The planners were appalled. They berated the planning director for even considering such blatantly political overtones from the chairman of the reapportionment committee. The veteran planners argued that the only variables to be used should be population figures and the principle of "one man, one vote." They argued that there should be no public hearings or meetings on this matter since it was to be determined on solely technical grounds. One of the younger planners asked if such a problem actually offered an opportunity to do some favors for the officeholders which would be returned someday when needed by the planners. The young man was soundly denounced by his colleagues for such an "unprofessional" attitude: it was suspected that his attitude was due to his failure to complete planning school.

The reapportionment map was drawn using data from the census and a standard program for creating uniform districts based upon population information. This map was technically so correct that the "one man, one vote" principle was guaranteed by an error of less than one-tenth of one percent for the entire metropolitan area. The project required three months of work and some $100,000.

The officeholders, joint reapportionment committee, minority groups, and public interest groups were incensed. Incredibly the technically correct reapportionment plan offended everyone. Within one week after its dissemination, this planners' proposal was discarded. In its place were several proposals drawn by officeholders, minority groups, public interest groups, and political parties. These were presented and debated before the joint reapportionment committee. The committee heard the proposals and closeted itself for one week to synthesize a new reapportionment map. It used retention of office, expansion of black districts, and political party affiliation and voting patterns as its three criteria. The plan was a reasonably sound political compromise that was acceptable to most groups. The federal court accepted the reapportionment, and it was approved by the involved federal agencies after relatively minor changes. Several suits were filed by disgruntled special interests, but all failed eventually. Reapportionment was accomplished successfully in this city and its suburban areas, and the city planners demonstrated their inability to deal with the political problems.

The Stadium Case

In a major city the regional planners were asked to find a proper site for a new major league football stadium to be financed with public funds. The planners were advised that downtown business interests were especially favorable to a central city location because it would enhance business opportunities and help boost the image of the city.

The planners established a series of technical advisory groups composed of various technicians from the highway department, local planning agencies, and other related public agencies. Curiously there were few politicians or businessmen on these advisory committees. The planners first addressed the question of whether there was a proven rationale for the new stadium and if it would be a wise expenditure of public resources given other needs of the city and its suburbs. They concluded that the stadium was a dubious undertaking and that the funds could be better spent on housing and health care. They recommended that if the stadium were to be built it should be modest in scale.

The planners and their technical advisers next established criteria for the stadium site in terms of traffic, surrounding land use, environmental effects, market area, and acreage needed. The planners then recommended three sites which were all in the suburbs some ten to twenty miles from the central city. These sites all met the technical criteria established, and the planners said that there were no suitable downtown sites available. Each of the three sites was adequate for a modest sized stadium.

The key businessmen of the city were most unhappy with this recommendation. They lobbied for building the stadium on the main artery into the downtown right next to the edge of major building concentrations. Doing so would necessitate clearing some black ghettos, but this could be financed with urban renewal funds, and concessions would be made to black politicians and businessmen to gain their support. The leadership of the suburban communities in which the sites were recommended by the planners were lukewarm to having the stadium in their backyards anyway. The planners, however, were entirely opposed to the recommendation of the business groups and launched a drive against the central city site by going on television, radio, and to the press in an effort to explain how the principles of sound city planning were violated.

Actually today when one views the grandiose stadium just as the central city rises in the immediate background, it is easy to see that the image of the city has been boosted.

Training and Experience Circumstances

A set of circumstances that contributes to further differences in the way that politicians and planners approach state and local problems has to do with their respective training and experience. These factors are manifested as well in professional attitudes and associations.

There is rather convincing evidence that the best students still go into such fields of study as law, business administration, medicine, engineering, and architecture, the more or less traditional professions. These fields in turn usually generate the politicians at the state and local level with lawyers and businessmen predominating. The pedagogical thread that runs through these professions is the reliance on case studies and accumulated knowledge. Case studies and accumulated knowledge are used as the foundation for professionalism and expansion of knowledge through research. This in turn allows for specialization within these professions. Thus today we have corporate, criminal, civil, and military lawyers, as well as tax accountants, corporate planners, dermatologists, internists, civil engineers, structural engineers, acoustical engineers, urban designers, landscape architects, industrial architects, and so on. Specialization within all of these professions is a fact of life.

State and local politicians in many cases retain their professional affiliation while holding office. Even in the largest cities and at the state level, the politician will maintain a continuing interest in and linkage with the profession for which he was trained. Often this entails membership in professional organizations such as the American Bar Association, American Medical Association, American Society of Civil Engineers, American Institute of Architects, and the myriad of business groups such as the American Manufacturer's Association, Society for the Advancement of Management, and the like. The professional linages of state and local politicians are strong, and the professional societies to which they belong are strong politically.

By virtue of being an elected officer in state or local government, the politico is eligible for membership in a host of political societies in addition to his political party affiliation. The National League of Cities and United State Conference of Mayors are important groups for mayors and top officials in city government. The Council of State Governments and Governor's Conference are significant for state chief executives. A corresponding group at the county level is the National Association of County Officers, and even the regional orientations of some politicians can be serviced through the National Association of Regional Councils. These groups are part political and part public interest, yet they all offer a source of information and support to members.

Circumstances of training and experience with its related aspects of professionalism are murkier for state and local planners. The usual practice for a person desiring to become a professional planner is to study a social science or architecture or engineering as an undergraduate field and then enter a master's degree program in urban and regional planning.[8] The basic educational principle for professional planning has been that graduate training was necessary for instilling the principles of urban and regional problem-solving (usually a two-year course of study).[9] There are of course numerous exceptions among planners with undergraduate degrees in planning (a few schools offer such degrees), undergraduate degrees with no graduate training or incomplete graduate training, and even some cases where professional planners have no university degree.

Curricula vary mightily among the schools of planning; emphases change; and faculty are diverse. Related fields such as economics, public administration, political science, urban studies, civil engineering, landscape architecture, and architecture often lay claim to being the real educational bases for planners. Thus the reality of the situation is that a person with a master's degree in urban and regional planning will undoubtedly have had different educational experiences than a person with the same degree from a different university. Furthermore, there is little interest or movement for standardizing planning education primarily because there is so much disagreement on what it should be. The diversity is desireable to many in that it offers the student a choice of what kind of planning he chooses to study, assuming that he is well-informed about the options available to him.

In recent years students have expressed considerable discontent with planning education programs because of their feeling that these courses deal with philosophical and methodological matters having little applicability or relevance to the tasks assigned a professional planner. Planning educators have responded in three ways. One approach has been to do nothing and retain the traditional curriculum. Another approach has been to literally abolish the curriculum and have a two-year program of electives, thus enabling the student to design his own version of a planning education. A third and most popular approach has been to develop curricula based upon a core of courses that allow the student to become a generalist planner and then allow him to choose a group of electives in a specialized area, thus the generalist with a specialty. Despite these sincere and energetic efforts, there is still little agreement among the educators, professional planners, and related professionals on what constitutes an education for professional planning.[10]

The chaos in planning education is symptomatic of a larger problem of defining what a planner is. Related technical professions such as architecture, landscape architecture, and civil engineering have been expanding their frame of activities to include more of the things that people calling themselves planners have been doing in the past. Social scientists are expanding their fields of application to deal more with state and local problems, especially within the rubric of urban affairs, which have belonged in the past to planners. Such diverse fields as business administration, medicine, and criminology are expanding the definition of their respective interests to include many aspects of what has been called urban and regional planning.

The American Institute of Planners is still the only professional society that lays claim to representing professional planners. In recent years the Institute has eased entrance requirements to allow people who have not entered the field in the usual way or who have not practiced planning in the narrow sense of land use and physical planning to become members. This has served as a minimal appeasement to related professions but has not solved the problems of definition of the field of planning. At a recent meeting of the Institute, a young man who identified himself as a socialist who had been trained and experienced in planning in the normal manner said that "the very ambiquity and softness of the planning profession and its institutional structures was what attracted me into the field since I can do anything I want and still consider it planning."

The Institute has been viewed as elitist by many related professionals, public interest groups, and politicians because its membership is small and restricted. The American Society of Planning Officials sought to provide a forum for information dissemination and interchange among an unrestricted membership that included professional planners, politicians, and interested citizens. It has been able to abate the discontent to some degree but has not been an organization that actively sought to lobby for and elevate the profession of planning if such activities would antagonize members who were not professionals. The Society has been more concerned with research, information, and increasing the opportunities for planning to take place–all necessary functions that the professional planners had not accomplished in their Institute. The irony of the situation, however, is that the membership of the Institute and that of the Society overlap, with the latter being slightly larger because of its organizational and citizen members.

Together the Society and Institute have not been able to clearly stake the boundaries of the profession or to emerge as a strong lobbying force. Curiously in recent years the leaders of the Institute have undertaken considerable efforts to foster special interest bills in Congress which would benefit at least the economic status of professional planners, and they have been taken to task for indulging in such nonprofessional matters by several of the members.

The basic question still confronting the profession of urban and regional planning is, Who should be the planner? One observer has portrayed the professional planner of the master's degree, American Institute of Planners variety as a modern day D'Artagnan at the top of the castle stairs sword-fighting simultaneously with architects, landscape architects, engineers, social scientists, and lawyers to defend the turf; and one can envision the politician on the top wall pouring burning oil on the planner.[11] Another observer-participant argues that planners may be able to identify themselves, but "one very real handicap to planning is that we don't know how to do it very well–We'll be alright if we know that we don't know all we think we know about it."[12]

Thus a summary of the differences between the training and experience of politicians and planners shows the dichotomy of the pragmatic individualist in a well-defined profession and the generalist groupie in an ill-defined and debatable profession. The planner has been under attack for several years by his friends and enemies for being poorly trained, inefficient, politically incompetent, and (from analysis of his experience) relatively ineffectual. Many of these charges have taken form in the bureaucratic wars for planning responsibility which transcend political and professional considerations.

The Tripartite Case

State planning in a midwestern state had been an important part of government for a quarter century. It included not only large area planning but local planning assistance as well. The staff was composed almost entirely of professional planners trained and with experience in traditional planning.

The agency had taken a particularly physical orientation and dealt in the main with matters of land use, natural resources, and recreation. There had been some critical comment that, even assuming the limitation of state and regional planning (and local planning assistance) to physical planning was valid, the agency should expand its interests to include the most important factors of physical development which are transportation, public utilities, and educational facilities. This comment was acceptable to many of the planners, but they felt it would take several years to develop a capability to handle such complicated matters while the highway department, public utilities commission, and education department were doing a reasonable job at present.

The state planning agency had survived changes in political leadership by divorcing itself completely from the governor and legislature, by being buried in a line department, and by having every employee covered by regular civil service classifications. The agency had been energetically turning out plans for state, regional, and local development for years with the aid of generous federal grants. The shop hummed along regardless of who was in power. While almost none of the planning recommendations were ever actuated or constructed, the professional planners felt that they had a good thing going and were adding more and more planners to the staff while raising the salaries of existing personnel.

A new governor expressed major concern about protecting the natural environment of the state and its subdivisions while at the same time improving transportation and economic opportunities. He wanted the state administration to act more efficiently and economically as well. Could it be done?

A reorganization commission was convened by the governor to examine these questions. After analyzing every state agency, the commission staff, which was composed largely of public administrators, business executives, and university professors, concluded that the interests of the governor could be achieved through reorganization and staff changes. The basic recommendation related to planning was that the existing planners were only concerned with land use and local planning assistance ("playing consultant" as one member put it). If this was what planners were interested in, then a division of local and regional planning should be established in a community affairs department and the staff reduced in size to handle a smaller workload. Transportation, public utilities, and education should be left in the existing agencies, but a division of the resources development department should plan for natural resources and other environmental factors. Finally there was a need for a central planning agency in a department of administration to relate the state planning activities to budgetary processes and the governor's programs. This central agency would coordinate all planning at the state and regional level.

The professional planners in the state planning agency were incensed at this recommendation. They expressed their belief that the reorganization commission had taken an extremely narrow view of professional planning and had attempted to relegate it to little more than land use planning. The planners argued that they had been trained as comprehensive planners who were able to perform all of the functions that had been defined in the tripartite organization recommended by the reorganization commission. The governor was not convinced so the planners took their case to the legislature which had to approve the reorganization. They lost.

The tripartite reorganization went into effect. The planners were placed in the regional and local planning division and restricted to providing technical assistance on land use planning to the various regions and local governments requesting such assistance. A group of engineers, naturalists, and assorted environmentalists were assembled and recruited for the environmental planning division of the resources development department. The central planning agency emerged as the real power source due to its super coordinative and budgetary responsibilities. This agency was staffed with persons trained and experienced as accountants and business executives, and with a few political supporters of the governor who were basically professional politicians. The state chapter of the American Institute of Planners protested most vigorously.

The Licensing Case

What is a profession and when should it be controlled? Many scholars and practitioners believe that only basic professions which are characterized by specialized training and experience, a code of professional ethics and conduct, and a body of knowledge should be licensed and even then only when there is a danger to the public from unprofessional practices. This has been construed to mean that architects, engineers, physicians, dentists, and lawyers are licensed everywhere. Yet many states license all sorts of professions from accountants to zoologists.

A group of consulting city planners in a large eastern state were upset by the entry of civil engineers, architects, landscape architects, and management consultants into the practice of certain aspects of urban and regional planning. Not only was this a drain on the limited funds available for consultants, but it constituted a challenge to the years of graduate study and yeoman internships that most of the professional planners had put in which qualified them as American Institute of Planners members. A vanguard of planning consultants quietly drew up a bill for submission to the state legislature which licensed planners and made it illegal for anyone other than a planner to prepare plans for a community. The qualifications for being a licensed planner were essentially those of Institute membership, although there was a "grandfather clause" as a concession to old-timers who predated many of the planning education programs. The licensed planners were enfranchised to develop land use, transportation, public facilities, and capital budget plans for local governments. Other professions would be allowed to perform their usual services such as street and highway engineering, architectural design of buildings, landscaping, and efficiency studies.

The bill was introduced lackadaisically by a state senator who was a friend of one of the consultants. The consultant himself was then president of the state chapter of the Institute. All up to this point had been so quiet that everyone was surprised by the silence. The head of the legislative committee to which the bill was referred happened to be a civil engineer, however. He estimated that there were only a few hundred persons qualified to be planners under this proposed legislation while there were tens of thousands of architects and engineers already licensed in the state. He called in the architectural and engineering professional societies. They were incensed and said that what was needed instead was a bill that would lessen the competition from planners into the fields of architecture and engineering referred to in the bill under advisement. After all, they argued, urban and regional planning is only a part of architecture and engineering. The bill that emerged from committee and was passed by the legislature said that any licensed architect, engineer, or surveyor could practice urban and regional planning in the state by simple payment of a fee for a license while any other person would have to prove his qualifications in terms of education and experience requirements established by a licensing board and then submit for an intensive professional examination. If successful he would then be allowed to apply for a planners license. Ironically, in their haste to endorse engineers as urban and regional planners, the framers of the bill neglected to define the various fields of engineering and hence a nuclear engineer or chemical engineer is qualified automatically as an urban and regional planner in the state upon payment of his fee.

The state chapter of the American Institute of Planners filed suit claiming that they had been deprived of their livelihood. The court trial centered upon many witnesses who attempted to show that urban and regional planners had skills for coordination and comprehensive linking of various aspects of a community which architects, engineers, and surveyors did not necessarily have without special training and experience. A lower court accepted the Institute's arguments, but a higher court reversed the decision and said such definitions of professional practice were within the prerogative of the state legislature which had acted with due process. The licensing law was upheld and still stands. The president of the state chapter of the Institute said that this was a shame since architects and engineers were "incompetent" to do urban and regional planning. The remark was widely published by the newspapers in this state. The professional societies of the architects and engineers promptly sued the state chapter of the American Institute of Planners for slander and libel, but the case was withdrawn when the Institute's state office issued a public apology.

The Maverick Case

Earlier it was stated that there are obviously bound to be exceptions to the general discussion that follows such matters as personality, education, training, and experience. An interesting case study of a planner who was quite unlike his colleagues occurred in the northeast. The planner did not go through the graduate studies routine into planning but instead entered the field from law school. He worked in several cities and then gained recognition in two cities in New England.

The planner's area of activity had to do with urban development and redevelopment from the public corporation stance. He learned early that the planner must at times act like a politician if he is to have meaningful successes. In his first major position he was somewhat ruthless in the manner in which he chastised critics and rewarded friends. Coalitions of minority groups and special interests were formed under his guidance. There were several cases in which those with the least political equity literally were squashed in the rush to implementation. In just a few years his efforts were reflected in the concrete and glass structures that stood as monuments to uncommonly effective planning.

In moving on to his second major undertaking, this planner was carried away with the heady wine of success and fame. He was making things happen and getting things done to the wonder and amazement of his colleagues, observers, friends, and enemies. He pursued his programs of planning and implementation with a zeal equivalent to the tycoons of the conglomerate set. His flaw was that he was highly visible and a likely target for advocates of the groups with the least political power and poorest circumstances.

In time this planner became an outcast because of his success among his colleagues in not only planning but related fields as well. He was a constant target for demonstrations and attacks on his inhumanity, coldness, and lack of compassion. His name and image became synomous with the detrimental effects of big government hurting the little man in the name of progress. The planner in question took all of this stoically and was somewhat bemused at how quickly one can go from celebrity to villain. Ironically this planner was very popular with several of the key politicians in the country and moved onto a very rewarding and significant position in state government.

This case study is included primarily as an illustration of a maverick planner who is successful yet is so different from his peers that he is chastized. Peer group recognition is important to planners, and the planner who becomes a politically successful activist is accused of no longer being a professional planner. He becomes suspect by special interest groups and an object of curiosity to many politicos.

Recently the subject of this case study returned to the city where he had made most of his plans come to fruition and addressed a major planning conference. He stated that perhaps he was not a planner anymore because of the way he had done things and the way most planners operated, but nevertheless, he said, "planning must be implemented through a political process and it was insignificant if it could not be so." He acknowledged that such an activist approach as his was likely to leave a trail of enemies and controversy, but this was the price of effectiveness. He said that he was optimistic that the professional planners were capable of reaching high levels of effectiveness and ability to handle complex political and social problems. His address was received with mild applause and politeness by the planners assembled.

At the end of this address, a group of planning students staged a demonstration meant to embarrass this planner. They assembled a collection of students and elderly ladies who were displaced by a redevelopment project initiated during his tenure (but not started until years after his departure). The demonstrators denounced his "inhumanly bad planning" as the cause for everything from inflation to racial segregation to elderly and student discriminatory practices. The demonstrators accused him of being power mad and despotic and an example of "everything that is wrong with planning." Many of the planners there cheered for the demonstrators.

NOTES

1. One of the first observers of the emerging split between value-neutral professionals and politically oriented activists in bureaucracy was Herbert Kaufman in his article "Emerging Conflicts in the Doctrine in Public Administration," American Political Science Review, Vol. 50 (December, 1956), pp. 1057-1073.

2. There have been several empirical studies of this dilemma of understanding. Among the more well known examples are: Francine F. Rabinovitz, *City Politics and Planning* (New York: Atherton, 1969); David C. Ranney, *Planning and Politics in the Metropolis* (Columbus, Ohio: Charles Merrill, 1969); Alan Altshuler, *The City Planning Process: A Political Analysis* (Ithaca, N.Y.: Cornell University Press, 1965); and Alan W. Steiss, "Planning and Decision-Making: Structural and Contextual Factors" (Ph.D. dissertation, University of Wisconsin, 1971).

3. Some of the classic community power studies are: Floyd Hunter, *Community Power Structure* (Chapel Hill: University of North Carolina Press, 1953); Roscoe C. Martin, *Grass Roots: Rural Democracy in America* (New York: Harper and Row, 1957); and Robert Dahl, *Who Governs?* (New Haven, Conn.: Yale University Press, 1961).

4. For an in-depth examination of the concept of satisfaction for decision-making, see Herbert A. Simon, *Models of Man: Social and Rational* (New York: Wiley, 1957).

5. The basis for much of the theory of the nonplace urban realm can be traced to Melvin M. Webber, "Urban Place and Nonplace Urban Realm," in Melvin M. Webber et. al., *Explorations into Urban Structure* (Philadelphia: University of Pennsylvania Press, 1964).

6. This point is presented convincingly by William Alonso in his article "Cities and City Planners" in *Taming Megalopolis,* ed. H. W. Eldredge (New York: Praeger, 1967), pp. 579-596.

7. The basic argument was made by Paul Davidoff in his article "Advocacy and Pluralism in Planning," Journal of the American Institute of Planners, Vol. 31 (November, 1965).

8. Master's degrees in urban and regional planning are offered at almost 70 universities, and related master's degree programs are offered at more than 60 universities. See American Society of Planning Officials, *Education and Career Information for Planning–1973* (Chicago: The Society, 1973).

9. For a thorough review of planning education in its classical and basic sense, see Harvey Perloff, *Education for Planning* (Baltimore: John Hopkins University Press, 1957).

10. For a dedicated effort to bring a minimal semblance of order to the educational chaos, see Ad Hoc Committee on Planning Education, *Final Report* (Washington: American Institute of Planners, 1973).

11. The analogy is from Richard F. Babcock, *The Zoning Game: Municipal Practices and Policies* (Madison: University of Wisconsin Press, 1966), esp. Chap. 4.

12. Frederick H. Bair, Jr., *Planning Cities* (Chicago: American Society of Planning Officials, 1970), p. 47.

Chapter IV

ON THE PEOPLE BEING SERVED

> *"If there ever could be a society of perfect men, there might well be as much competition to evade office as there now is to gain it; and it would then be clearly seen that the genuine ruler's nature is to seek only the advantage of the subject, with the consequence that any man of understanding would soon have another to do the best for that other himself."*
>
> –Plato
> *The Republic*

On one of the immortal Amos and Andy television shows, Andy is seen entering the Kingfish's room and is excited about the proposed urban redevelopment program for their neighborhood.

"Wha da y'all think uh dis heah 'city plan,' Kingfish," Andy blurts out.

The Kingfish puts his thumb and forefinger to his chin and pensively strokes an imaginary goatee, then peers over his horn rim glasses and pronounces, "Well dat dere depend upon wheduh yo da planner o'da plannee."

The humor of Kingfish is indicative of more serious considerations affecting people with regard to planning and politics. The almost rhetorical question is: "Planning for whom?" One could argue as well: "Political process for whom?" Such questions lie at the heart of not only planning technique and methodology but the very essence of democracy and republicanism. The standing response that "planning is for people" is insufficient per se. It gives rise to only a minor improvement in specificity when the argument is extended to include: "Planning is for the public good." As has been stated previously, the public good (or planning for the people) is a euphemism for saying that, in reality, the "plannees" are illdefined. Such a euphemism nonetheless eases the burden of proof for both planners and politicians but has tended to yield woefully unsatisfactory results. If planners and politicians cannot or will not be specific about the people being served, then such planning and implementation will guarantee ineffectiveness. If it becomes necessary to include elements of compromise and balance of special interests in order to make planning and politics acceptable to specified people, then so be it if it works.

Pluralism, Participation, and Planning

Life in the United States is characterized by a pluralist nature which pervades not only politics but the economic, social, cultural, and spiritual fabric of the nation. Arguments to the contrary are not convincing or realistic. The political process that has evolved is related directly to pluralism and allows differing interests to foster their own wants and needs. Since this political process is rooted in a limited democracy rather than a pure democracy, participation is restricted somewhat. The organization for this political process is the republic form so that representatives of the people foster their needs and wants—not a perfect organization but an operational format. All of these factors have combined to create a need for participatory planning whereby representatives of the people, special interest groups, planners, and often key individuals reason together. This is different from an idealized state in which a technical elite makes decisions for the uninformed masses, even though such a scenario is attractive to any mortal engaged in planning, and it forces planners to deal with politicians, special interests, and key individuals, whether they like it or not.

Participatory planning, like its larger context of participatory democracy, requires an understanding and ability to work with issues of political influence. The manner in which special interest groups (in their broadest sense) form, grow, and coalesce with compatible interest groups is required knowledge for planning.[1] Such knowledge on the part of the planner enables an improved basis for not just doing planning but for effectiveness in determining public sentiments and transmitting these in the form of influence for planning proposals to elected representatives of the public.[2] This is not a usurpation of the politician's responsibility but an aid to helping the politician meet his responsibility.

Participatory democracy has been intertwined with citizen control over public affairs throughout the history of the United States.[3] The early town meetings in New England laid the foundation for representative government later established in Virginia and the other twelve colonies which was unlike anything known in Athens or London. The liberalizing of qualifications for citizenship and the entire spectrum of the Jacksonian Revolution were major factors in the extension of the control of public affairs to more people. These beginnings set the tone for the contemporary set of voluntary organizations and formal associations, as well as primary organizations (family and friends most notably), which comprise special interest groups for state and local political life. This thread of participatory democracy has grown to mean a process whereby common amateurs of a community exercise power over decisions related to the general affairs of the community.[4] Indeed, as Max Weber argued, such a process is essential if a community is to survive.[5]

The response made to the need for participatory planning has been significant only since the early 1960s. The period from 1890 to 1960 could be characterized as one in which the planner was left to his own devices to include citizen involvement in making recommendations for plans and implementation. This largely took the form of utilizing the independent planning commission as the most convenient and accessible point of contact with citizens. An overreaction set in during the mid-sixties in the form of somewhat disastrous experiments, largely through poverty elimination programs sponsored by the federal government, which attempted to bridge a gap of seventy years in a few months. The misguided and misinterpreted dogma of maximum feasible participation of groups most affected by the programs being planned was abused sufficiently by minor demigods to create not only worthless and infeasible plans but also a backlash that almost led to autocracy and minimum feasible participation. The pattern has moderated in recent years to include orderly and well-structured mechanisms to attain participatory planning through direct and indirect citizen participation.

Distinguishable patterns of citizen participation can be examined for each of the three forms of planning that were defined previously. The basic patterns of the Technocratic Form have been those which garnered the greatest amount of criticism. Not all of this criticism is valid, and much of it is cheap shots at a vulnerable profession. Often-heard complaints that planners are "trained to feel superior" and "employed by superior powers not by the people he should serve" are obviously drivel.[6] The more substantial criticisms of the pattern of participatory planning used in the Technocratic Form have to do not with a conspiracy of elitist schools and governments but rather with the structure of this classical model.

The Technocratic Form of planning has relied in the main upon the independent planning commission to orchestrate the involvement of citizens. Four basic features are involved.[7] First, the independent planning commission is assumed to be representative of the entire community and guardian of the whole public interest. Second, the independent planning commission is assumed to be capable and responsible for defining the long-term goals and short-term objectives of the community from which plans can be devised. Third, the plans that are devised usually take into account physical aspects and facilities of the community. And fourth, the day-to-day decisions on projects and developments for which the independent planning commission advises the government are measured always against the long-term goals. This structure is supposed to fix the whole public interest into all decisions and hence enables a whole public rationality.

The critics of this structure with its four salient features have had little difficulty in demonstrating its futility and ineffectiveness.[8] The critics usually show that the independent planning commission is not representative of the pluralism of the community nor does it adequately represent major special interests. On the other hand, the architectural, real estate, and engineering special interest groups are overrepresented commonly. There is little evidence that any independent planning commission anywhere in the country has ever defined the long-term goals of its community in any greater detail than the need to cherish motherhood and preserve hot dogs and apple pie in the American diet (even these goals have come under attack). Rather than being a technical problem, this is more a matter of an incomplete understanding of the way to mesh special interests of the pluralist community primarily because the independent planning commission is incompetent to perform such a function on its own. It has been shown time and time again that the independent planning commission and its professional staff have been able to influence decisions to only a limited degree; and hence, even if the citizens were involved in the planning process, their needs and wants might not be incorporated in its decisions. The assumption that long-term goals which serve as the basis for long-term plans will be used as the guiding framework for the advice given to the government by the independent planning commission becomes dubious when such factors as influence, prestige, equity, and power for competing groups are considered. Other criticisms of the Technocratic Form could be detailed, and in fact it is fairly easy to demonstrate such examples of failure. The failure has resulted from naive assumptions and poor structures for citizen participation, however, and critics who charge that planners are pawns in a plot by the conservative upper class to maintain the status quo are off the track and counterproductive.

The problems that have arisen concerning the pattern of citizen participation common to the Technocratic Form have been an inability to relate to any special interest groups consistently or well. Such problems tend to frustrate the members of the independent planning commission so that they lose interest in their responsibilities and become apathetic. The professional planners are equally frustrated and tend to dwell upon the institutionalized aspects of their work in a manner that elevates the importance of technical matters to unrealistic heights and leads to identification of the staff as its own client. Such a curious set of relationships for planners has its defenders who argue that planning as an institutionalized activity of government should strive for technical perfection in order to satisfy other professional planners on the staff and elsewhere as the basic client group and then present the results to the independent planning commission as the second major client group.[9] This absence of true citizen participation often leads to practices that are essentially in the self-interest of the planning staff and independent planning commission in such a manner that any citizen participation becomes undesirable except when it reflects these self-interests.[10]

The Participatory Form has used the concepts of advocacy planning to overcome many of these problems and deal with distinguishable special interest groups. Aside from the problems of mixing ideology with the advocacy planning technique and the resultant domination of the efforts toward special interest groups with the least political equity as mentioned previously, the Participatory Form nonetheless has made some improvements upon the classical features of the Technocratic Form for citizen participation.

The Participatory Form rejects the notion that the independent planning commission or its professional staff or any similar groups can define the whole public interest or determine the best plan from among several alternatives. While not challenging the political process and its weighing of special interests, the advocate planners seek to better include all interested special interest groups. There are four basic features for citizen participation in the Participatory Form as typified by advocacy planning.[11] First, the nature of pluralist interests represented by various special interest groups for the community is accepted. Second, it is assumed that the interested groups in planning will be represented by professional planners as advocates for their special interests. Third, it is assumed that each group or coalition of groups will present its version of the plan to a planning commission or, in its absence, directly to the governing body. And fourth, the forum will select a plan or compose a plan from the various alternatives presented. The Participatory Form sees this set of features as sufficient and necessary for viable citizen participation as well as guarantees for the use of the articulated needs and wants.

For reasons quite different from those given for the failure of the Technocratic Form, the Participatory Form, with its use of advocacy planning, has had relatively little success in improving citizen participation in planning. The basic problems have been that the special interests groups do not have access to professional planners as assumed; there has been little financial support for the advocate planners from government; and the structure of the process takes too much time for most planning problems.[12] The obsession of the advocate planners with minority, poverty, and radical groups has created image problems with many state and local politicians, but even more problematic has been the tendency of advocacy planning to serve more as a reaction to plans rather than a formulative endeavor.[13] There are several cases where advocate planners have not been able to prevent a plan they believe to be inimical to their interests from being adopted, and they have resorted to the courts with mixed successes.

The recent posture of the Participatory Form planners has been that, while their brand of citizen participation has not changed the world, they have nonetheless made a start. The early zeal and radical tendencies of some of the advocate planners are waning, and there are even some signs of limited institutionalization and professionalization of the movement which tends to stifle major risk-taking.[14] Thus the efforts of the advocate planners to use the Participatory Form have been most interesting but not of sufficient significance to consider the movement as substantive and long lasting.

The Activist Form has not stressed citizen participation in its previously stated definition both because of its recent arrival on the planning scene and because of its inherent linkages with elected officials. Instead, the essence of the citizen participation features associated with the Activist Form is the implicit admission that politicians can do a better job at it than professional planners.[15]

The basic four features of citizen participation in the Activist Form are as follows. First, the politician is regarded as a representative of various special interest groups rather than of a hypothetical public interest. Second, the special interest groups can go directly to the politician to foster their needs and wants rather than going through the intervening level set by planners. Third, the planners can help articulate and synthesize the needs and wants of the special interest groups in the community and strive to ensure fair and equitable presentation of views. And fourth, the decisions will be rational because the politician must satisfy the needs and wants of the special interest groups of the community if he wants to survive in politics. The nature of citizen participation used by the emerging group of the Activist Form is that of the conductor of participation rather than the doer.

To be sure, there has been little experience with this new twist in citizen participation by the Activist Form because there are so few professional planners who are engaged in this type of activity in the main. In this sense the emerging Activist Form, like the Participatory Form, has not had a great impact upon planning at the state and local level. Some critics argue that this approach is limited because in practice only the most powerful special interest groups are able to reach the politicians.[16] Yet there have been considerable signs in recent years that indicate gains made by community organizers and politicization among various groups which seem to make this criticism archaic. The elegant simplicity of the direct access of the special interest groups and planners to the politician underlies the beauty of the Activist Form and is the key to its great potential for improving citizen participation.

The Seashore Town Case

The classical approach to providing citizen participation in planning, usually associated with the Technocratic Form as discussed herein, has been to conduct public hearings on planning matters. Most states have enabling acts which require public hearings to be held on most planning matters and hence this approach to citizen participation is institutionalized in law to such a degree that it will be with us for many years to come. An interesting case study of the medium-sized city on the coast of the northeastern state sheds some light on the effectiveness of the public hearing as a conveyor of public sentiments.

The city had been awarded a federal grant to complete a plan which dealt with such matters as land use, transportation, housing, and capital spending. Conditions of the grant stipulated that the city and its planners should make concerted efforts to obtain citizen participation in the planning process. The city retained a planning consulting firm to accomplish the technical work, and the independent planning commission was assigned the responsibility of coordinating the work. State law required that a public hearing be held to assess public sentiment before the plan could be adopted by the city.

The independent planning commission was composed of nine members, among them two architects, two engineers, two persons involved in real estate, a businessman, a preacher, and a housewife. The members of the commission believed that they represented the community and were fairly competent to reflect the needs and wants of the citizens. Opportunities were afforded for citizen participation through open meetings concerning the plan which were widely publicized. Very few people came to these meetings, and it was sometimes difficult to get a quorum among the commission members.

The consultants performed the technical aspects of the planning in a capable manner, but they turned constantly to the members of the commission for advice on goals for planning. There was neither the time nor the money for the consultants to pursue more sophisticated means of citizen sentiment determination. This fact, coupled with an underlying queasiness on the part of the commission members about turning the outside consultants loose on the community, resulted in almost no direct involvement of the polity. The city council, which was responsible for the ultimate approval of the plan, was aware of the progress of the planning efforts and offered concurrence in the approach.

The substantive nature of the plan was such that an overall goal of tourist-based growth was advanced. Older, low-income housing areas which were close to the shoreline were to be replaced by tourist accommodations and some public beaches which would be operated by the city. Fees would be charged to nonresidents for use of the beaches, and various franchises and concessions would be made by the city to uplift the character of the vacation-oriented commercial enterprises. Highway improvements would be made to facilitate the ingress and egress of vacationers. A capital improvements program was proposed which used combinations of general obligation and revenue bonds to finance the plan, and recommendations were made for further use of federal grants. Given the goals defined by the independent planning commission and implicity approved by the city council, the plan was rational and stood a fairly high chance of being successful.

The city council and independent planning commission decided to hold a joint public hearing to meet state statutory requirements prior to adopting the plan. The statute called for publication of the substance of the plan in a newspaper of general circulation in the community before the public hearing.

Overflow crowds packed the city hall auditorium, and a nasty and angry air filled the room. It was evident early in the evening that something had been brewing of which apparently the independent planning commission and city council members had not been made aware. Speaker after speaker condemned the plan as commercialist exploitation and implied all sorts of payoffs and conspiracies by the politicians, planners, and business interests in foisting this plan upon the community. They talked of a quality environment, control of growth, peace and tranquility of residential areas, and preservation of the values they cherished in this community. Most of the speakers were representatives of hastily formed neighborhood organizations or they appeared as private citizens. In general they tended to be long-time residents of the city. A few members of the business groups spoke in favor of the plan as did several members of the state and county economic development agencies. After five hours of testimony, however, it became quite clear to the planning consultants that some severe cleavages existed between what they were led to believe were goals for the future of the community that were reflective of the common interest and what had been said that evening. The meeting was adjourned on several angry notes and the threat of court fights in the early morning hours.

It was the custom of the city council to take a vote after public sections of the hearings were closed. The chief planning consultant apologized for the errors made in assessing the goals of the community and stoically accepted the blame for the misinterpretation and in doing so exonerated the independent planning commission.

The mayor of the city looked puzzled and said: "What are you talking about, son? The plan is just fine and just what we wanted."

The chairman of the planning commission then chimed: "You certainly should be used to this sort of adverse reaction from a bunch of malcontents. They certainly don't represent the real community."

Upon the advice of the planning commission, the city council then proceeded to adopt the proposed plan. The chief planning consultant offered his further services for modifying the plan, but was advised that it was not necessary to change the plan in any way.

Somewhat depressed and bemused, the consultants left town and made the long journey back to the home offices.

They had to get some sleep because work started on a new plan in a larger city the next day.

Small groups of neighbors and friends talked for many weeks about the apparent evils of the planning process and the conspiracies that existed between politicians, planners, and greedy profit seekers. Several talked of court suits, but few had the money to donate or the confidence that anything could be done. They figured that elections would be held in a few years, and there was little likelihood of the plans being implemented anyway.

Cui Bono

More than a few repetitions of the Seashore Town Case in recent years have focused increased attention upon the questions of who gains from planning decisions. Related to the heightening of this movement are related efforts for consumerism and neoreformism. While there is much evidence to demonstrate that the silent majority is resigned to the depressing posture that selected special interest groups will always derive benefits from planning decisions, there are several groups that are the least vociferous if feeble. A general feeling of cynicism and suspicion nonetheless surrounds questions of who gets what from planning. The increasing effectiveness of minority and poverty special interests groups in wresting from planning programs direct benefits which go to a few members rather than to many only leads to further skepticism.

The persistent Devil Theory pervades much of the rhetoric and polemic. This theory holds that a power elite is in charge of things, and politicians and planners merely serve as their handmaidens.[17] This "power elite in every town and city in America" towers above the general populace and creates a conservative political structure immediately below it to preserve the status quo and allow the power and wealth of the power elite to grow.[18] Professional planners must serve the interests of the power elite and its political structure if they wish to survive in their jobs, and hence the planners do not strive to involve special interest groups in the planning process except when those special interests are power elite compatible. This theory holds that the Technocratic Form of planning especially is lacking in any ideology or ethics other than the preservation of the status quo and the enabling of growth that is in the interest of the power elite.

Such critiques of the planning practice in this country are popular and enjoyable. It is somewhat of an American pastime to bemoan the Devil Theory conspirators who are greedy and lustful and the cunning way they use planners to attain their ends. The popularity and fun of such theories does little to enhance their credibility, however, and few take them seriously except when they serve as useful points in a rhetorical argument. It is of course sheer nonsense to assume that some conspiracy has been perpetrated on an entire profession and that most elected officials are dupes of an aristocracy hidden and shielded from the public wrath. Not only are the organizational and supervisory elements of such a conspiracy beyond belief, but to think that a Devil Theory operates at such a scale with no exposure is naive. The Devil Theory of the power elite dominance of state and local politics and planning to the exclusion of all other special interest groups is readable and inspiring fiction but utter nonsense.

An alternate theory to explain the nonsystematic exclusion of many groups and individuals from the planning process, and probably the political process as well, is the Naive Theory. The Naive Theory applies more to planners than to politicians, but there are undoubtedly many instances where the politicians may know what is happening but remain silent for any number of reasons.

The Naive Theory holds that planners can be taken in by sharp special interest groups who know how to participate in the political process. These special interests elicit support from the politicians for one set of reasons and then attempt to gain the planners' support sometimes by downright deception based upon laudatory and intellectualized responses to planning techniques. By giving the appearance of cooperating fully with the planners and being in complete agreement with the methods needed for attaining the principles of planning, some special interest groups can so impress the planners that planners naively believe that little further citizen participation is needed. Organized opposition groups who appear to challenge the special interest group that has persuaded the politicians and planners are somehow branded as mavericks that are not to be taken seriously but are nevertheless allowed to play the game.

Another application of the Naive Theory is the somewhat vaguely defined area of insider knowledge of planning proposals. Regardless of the content of many planning proposals, an advanced knowledge of such things as the location of highways, zoning changes, rapid transit stations, new social services programs, and new capital projects is sufficient to make a great deal of money. The unearned profits or increments may be of such value that some special interest groups fully support the proposals in order to make their own gains. The planner often misinterprets the true reason for this support and instead basks in a temporary euphoria over having accomplished an ideal. The politicians may or may not be aware of the real reasons for some special interest group support, but often they see nothing wrong in somebody's making a profit from a project that they think is probably good for the community at large anyway. The Naive Theory in this scenario can take many forms, but the common result is that the planners come out of it looking as though they were in cahoots with the forces of avarice all along.

The Planned Community Case

A very interesting case study was recorded recently in the suburbs of a major southern city experiencing a very high growth rate. Middle-class white families had been leaving the central city for years for a number of reasons including the desire for more spacious play areas and lawns, as well as the nondesire to live near black families. The suburb in question had only a few thousand residents but a very large geographical area which could be expanded even further through annexation. The central city had been threatening to annex many of the unincorporated areas outside the suburban town in order to recapture some of the fleeing property tax base, but the proposal was incredibly unpopular in the state legislature which had to approve such a move.

A large and reputable developer in the area quietly assembled several thousand acres a few miles outside the suburban town and began plans for a planned community by hiring a good land planning consultant. The land was mostly comprised of small farmsteads and was zoned for agricultural use by the county. A new freeway leading to the central city, which would make the large site completely accessible to the entire region, was under construction.

Continuing the very quiet process and low profile, the developer approached the mayor of the town and revealed his plans, but he requested that the mayor keep everything secret in order to prevent rampant speculation. He told the mayor of the advantages of the plan that would include all needed public facilities and bring an incredible increase in the property tax base should the town wish to annex the area and approve the proper zoning. The developer implied that the county planners were not too favorable to the concept because it would require major efforts on their part and they were "anti-development" anyway. The developer said that he wanted the annexation especially not because of the more relaxed zoning ordinance but to protect his future residents from what he thought would be illegal forced integration of neighborhoods and schools perpretrated by the court system. The mayor was no dummy, but he found logic in much of the argument.

The mayor arranged a meeting with the developer and the town's planning consultant after meeting several more times with the developer in the region's finest restaurants and bars. The mayor told the planning consultant how much he liked the proposed planned community because it would not only guarantee the kind of good development he sought but would cost the town almost nothing since the developer would pay the facilities costs. The mayor asked the planner to meet with the land planning consultant to the developer and work out the details for the site plans and zoning and subdivision regulation strictures. Both planning consultants so proceeded, and the mayor and the developer began to woo the rest of the city council and planning commission.

By the time the plan was completed and announced, most of the preparatory work for annexation of the area and zoning revision had been established. The plan in technical terms was indeed a sound one and included many principles of planned community design from the literature of the fields of planning and architecture: it was a noteworthy technical endeavor. There was no reason to suspect the developer had any intention of not following through, and he stated on numerous occassions before the politicians and planning commission that he would see the project through its completion in his lifetime (the plan called for a twenty-year development program to house twenty-five thousand people and provide all of the appurtenant work, shopping, and social necessities). The planning commission and the city council annexed the area and changed the zoning map to allow for the development. The zoning change was to apartment and commercial land because there was no special category for planned communities, and no bonds or covenants were required because the developer was well known and had worked so closely with the town's planning commission and planning consultant. The only opposition to the entire state of affairs was a small and unorganized group of surrounding property owners who basically opposed any kind of development in the area; they were not at all effective as a special interest group. Many warnings were sounded from throughout the region, but the local politicians thought that these were just jealous neighbors or central city interests intent on picking the plum for themselves.

For the first year after the annexation and zoning revision, little happened with the land but the opening of the freeway had begun to affect land values. Ominous signs began to appear on the property suggesting that information could be gained by phoning the developer. Quiet sales of the property began at astonishingly high prices. The site for the major department store was sold to a retail gas service station. The site for the first-class restaurant was sold to a large hamburger chain. The first village area proposed in the plan was sold to a commercial apartment building firm. The pattern of piecemeal deterioration continued so that the planned community proposal eroded into a simple extension of suburban commercialism and apartment building. The developer at first defended his actions saying that the minor deviations were necessary to allow for money to be raised to finance the planned community which was now beset by inflation woes. Later on he simply refused to comment on the property sales other than saying that they were legal.

The politicians became somewhat curious and began a small-scale inquiry which revealed that all was legal under the terms agreed upon at the onset. The city attorney thought that the politicians had put too much trust in the developer, but there was little to be done about it. The politicians felt compelled to mildly defend the developer since failure to do so would be humiliating and dangerous to their political longevity.

In a few years the area was developed in a rather typical suburban sprawl of gas stations, fast food stores, apartments, and single family homes. The proposition was a losing one for the city which had to raise taxes several times to provide the services needed by the new residents. The older residents felt cheated and swindled and were convinced that the politicians and planning commission, including their professional planning consultant, had been bribed and otherwise bought by the developer.

The town's planning consultant later admitted privately that he had been naive to think that the developer could resist the temptation to make large, immediate profits in order to retain his integrity and respect with the local politicians. The consultant also admitted that the planned community proposal was probably used as a ruse by the developer.

One night in a bar, the consultant drank hard on his bourbon and then braced and said: "It was a hell of a good plan though."

Points of Intervention

In order to serve the people of a community, the political process as defined previously allows for certain points of input and direct participation. Traditional views of planning tend to instill a notion of the planner as the overall seer of the public interest, and hence the planner emerges as simply another special interest group in political terms. Recent views of planning as a service to people in a community are oriented toward planners as conductors of public involvement and public control of governmental decision-making. Such views require that planners identify clearly the special interest groups involved and ease the submission of input. The planner also may be able to assist in the articulation of special interest group views. This recent view of citizen participation and fostering of community sentiments is more operational than earlier views and stands a higher chance of success. The planners must change their strategies, however, to deal with the key points of intervention in the process by improving their communications and influence with community groups and their leaders. These points of intervention allow the planners opportunities for modifying and refining group goals if that becomes desirable.

Opinion Leaders

The planner should be able to better identify the special interest groups in a community and determine who represents these groups. The representatives of these groups can be called opinion leaders since they voice group interests as well as provide leadership in the formulation of group interests. Because planners have had a dubious record in accomplishing this linkage with opinion leaders, it becomes clear that some strategical adjustments are needed.

Identification of special interest groups is one of the most unstructured and vague problems for planners. Planners have shown evidence of picking vociferous opinion leaders of a liberal bent more than conservative opinion leaders to help articulate needs and wants; however, most planners realize that all sides of the argument should be heard under ideal circumstances.[19] Yet determining precisely who is the opinion leader of a given group is problematic for planners.

The opinion leaders can be the formal representative of a special interest group or the informal leader of the group. Most groups designate some person as chairman, president, or liaison. Such a person may or may not have authority to speak for the group on planning matters. In many cases the person best qualified to speak for the group may not have any title or badge of leadership other than the respect of the group. Thus a key to identifying opinion leaders is the element of respect which is linked with influence. Such factors taken in conjunction with the specific organizational style of a special interest group are useful for determining who are opinion leaders. There are no hard and fast rules, however, and the identification of special interest groups and opinion leaders is one of the arts of planning.

A particular point of intervention for special interest

groups is through the opinion leaders. Much research in recent years has shown that opinion leaders tend to affect both the group, which of course includes individuals who may or may not be members of the group, and the mass media aspects of opinions.[20] Sometimes called the Two-Step Flow Hypothesis (see Figure 3), this empirical theory holds that mass media opinions, such as editorials, news coverage in some instances, and policy statements, as well as political support policies, tend to be filtered through the opinion leaders before affecting the position of the group. Conversely the groups usually interact with the mass media through the opinion leaders. This tends to destroy the notion that atomized automatons sit in front of televisions and radios or read newspapers and magazines and believe everything they see and hear. While there are notable instances where the groups may express an opinion to the mass media in the form of group or individual letters to the editor, feedback responses, telephone calls, cables, or telegrams, the major flow of communications that affects opinion stances is through the opinion leaders. Even when the mass media reaches the individual members of the group, it usually does not lead to a concrete opinion or position by the group itself without the interaction with the opinion leaders. Thus the opinion leaders have a formidable role in the determination of group

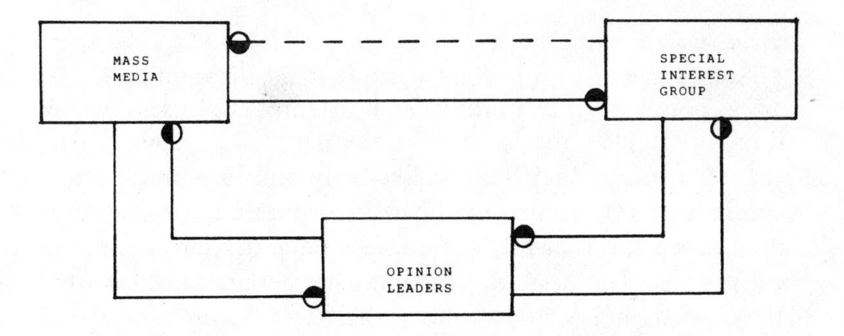

Figure 3. A VARIATION OF THE TWO-STEP FLOW HYPOTHESIS

opinions which become public sentiments in the aggregate. There are many styles of opinion leaders and hence whether the opinion leaders are reflecting opinions expressed to them or actively attempting to impose their opinions on the group is of somewhat peripheral interest to the planner. The planner can thus utilize the opinion leader as the key point of intervention for citizen participation through involvement of the various interest groups of the community.

Such a strategy obviously requires that the planner coordinate his work more fully with the elected political leadership of the community. In some instances the political leaders may indeed be of the opinion leaders, but in many instances opinion leaders may be adversaries of the politicians. This need not hamper the point of intervention strategy, but it places some constraints upon the scope of involvement. Such instances may require that a planner take a political stand or a stand bordering on a political factor in order to employ the strategy of points of intervention. This entails elements of risk for the planner to a sufficient extent that careful analysis should precede taking the risk. The reality of the situation, however, requires that the planner work closely with the political leaders for both their assistance in identifying opinion leaders and their insight and warnings.

Value Analysis

Sometimes it is important to go beyond the opinion leader as a point of intervention and attempt to assess values directly. If planning is to serve the people in a community, then the values they hold collectively and in a plural set of special interest groups should be incorporated into the planning process; it is clearly incorporated directly into the political process. The best way to find what people value, or what they need and want, or how they feel about matters of interest of planning is to ask them. Politics is the essence of such an interrogation of people, but planners can be valuable assistants to politicians in determining values, especially through the use of more or less technical devices.

It is not possible or effective to straightforwardly attempt to measure values. Such a complex measurement requires a technical assessment. It is possible to ask for opinions, however. An opinion can be a long-term, general thrust of thinking that affects decisions of the group. Such an opinion might be called an attitude. Or an opinion can be a short-term, specific reaction that may or may not affect decisions of the group. This type of opinion might be called a perception. Taken in concert, perceptions, attitudes, and opinions can be indicators of values of groups and their individual members. Determination of such values is mandatory for the articulation of community needs and wants as well as goals and objectives that are needed to do planning.

There is a small arsenal of devices for measuring opinions. The planner can use one or more of these devices for the measurement of opinions and then apply judgment and subjective analysis, as well as statistical inference, to arrive at conclusions. The final aggregation may not be what is directly adapted as goals and objectives for planning, but it is at least a valuable source of knowledge for both the planner and the politician.

There is a general classification of devices for opinion measurement called survey research. The most well known type of survey research is the opinion poll which is usually a quick sampling of people on a few matters. The scope and depth of opinion polls are constrained usually by limits of time and efficiency. The more in-depth and long-term version of survey research is called the attitude survey. The attitude survey often involves intensive interviews of individuals or representatives of groups in order to explore a series of perception, opinion, and attitude scales that can be used in a composite called an indicator series. A rarely used but interesting version of the opinion poll and attitude survey is borrowed from psychology and called the clinical survey. The clinical survey is composed essentially of relaxed discussions and observations of individuals who are carefully selected by virtue of how well they represent a special interest group or a set of characteristics that are common in the special interest group. With all three types of survey research there are inherent problems of a technical nature, but most of them are resolved according to the time, effort, and money available.

Content analysis is a set of techniques for analyzing the meaning of the various communications of special interest groups by analyzing patterns of words, sayings, expressions, and sentiments for consistency and regularity. These techniques often enable a kind of surrogate for survey research, assuming that there are sufficient examples of communications between and among groups and government and media; but they usually require the services of a statistician. A similar set of techniques is called decision analysis. Rather than concentrating on communications of groups, the decision analysis adherents argue that only the decisions that result are significant, and hence the analysis revolves around budgets, laws, regulations, guidelines, administrative rules, program specifications, and other examples of decisions. The decision analysis techniques can be valuable for understanding how group values are reflected directly or indirectly in the output of the political process. It can thus be extended to show that this is the most important aspect of planning implementation.

A couple of exotic techniques are of interest. Anthropological surveys are modern applications of the field for dealing with various kinds of groups usually in an urban context. Ethnic groups, for example, may exhibit certain patterns of thought and action that are indicative of values that will be expressed with regard to planning matters. Other groups with consistent patterns of thought and action, such as those expressed by religion, race, income, age, sex, politics, and education, may be surveyed using techniques of anthropology for the determination of opinions and values. A final technique of interest is market research, which is being transfered to the public sector. Market research can both define and identify the needs and wants for toothpaste and cigarettes as well as those for rapid transit and public housing. Hence there have been many recent examples of the application of market research techniques for a kind of community participation in government.

These devices can be useful for bridging the gap between the infeasible maximum feasible participation camp and the autocratic and technocratic camps. Some critics argue that this is a Big Brother subterfuge to avoid dealing directly with people in the community and argue for more personalized methods. The stark reality of the matter is that direct and highly interpersonal citizen participation is not possible in complex and large-scale problems. Rather than avoiding citizen participation, it appears that these techniques can be defended as raising the level of citizen participation in the determination of group needs and wants. The historical constraints upon citizen participation have been accessibility, time, money, and technology.[21] Given sufficient money and utilizing the technology already available, these devices can not only ease the constraints imposed by accessibility and timeliness, but they can be used by planners and politicians as well. Such devices enable the planner to perform more realistically and effectively as a conductor of citizen participation.

Does an improved capability to assess values of special interest groups mean that planning by polls becomes inevitable? Not necessarily, if the context of this discussion is maintained. That context is principally the validity and necessity of identifying the supports and demands that are used for the decision-making process of government. Hence it is not precisely a transference of opinions into decisions except in those instances where the burden of opinion may be so overwhelming that elected political leaders have no real choice but to make the transference. Within this same context it should be mentioned that the point of intervention strategy may be relevant also when there are clear issues that the planner and politician feel are not understood, are misinterpreted, or are otherwise deficient so that they are not manifested in values as indicated by opinions. This might be called the leadership potential for the point of intervention strategy.

The Model Cities Case

A rather unusual and interesting case study involving various aspects of citizen participation and points of intervention took place a few years ago in a major metropolitan center as part of the Model Cities Program. The Model Cities Program was an attempt to synchronize the expenditure of federal categorical grant programs for the poorest and most rundown neighborhood in major cities. It included a guideline provision that required maximum feasible participation by the residents of the neighborhood defined as the Model Cities Area.

The mayor and councilmen of the city felt that a black man should be the director of the program because the community designated as the site for the program was overwhelmingly composed of blacks. The new director, a black professional planner, met with several clergymen, businessmen, teachers, and politicians from the Model Cities Area. They decided that the best way to fulfill the guideline for citizen participation was to hold a mass convention of residents and let them elect a Board of Directors for the program. The Board would select various standing and ad hoc committees to prepare policies for the specific features of the program. Because the people would elect these directors, they would be participating fully in the program.

The mass convention attracted three hundred people. Forty thousand people were eligible to attend since the only requirement for voting was a residential address in the Model Cities Area. The largest share of the attendees seemed to be mothers with small children in tow. There were few men, and there was only a handful of high school and college age young people. The mass convention elected a fifty person Board, most of whom were in attendance. The Board was about half female (predominantly welfare mothers with a few working mothers). The remainder were clergymen, businessmen, and a few educators. Many of the Board members could be identified properly as opinion leaders in the Model Cities Area.

A fairly large staff was assembled with major emphasis placed upon hiring residents and blacks. Substantial federal funds were pumped into the program to enable the plans to be made. Most of the staff working on the planning aspects were not professional planners in the traditional sense but had worked in closely related areas such as housing and welfare programs; the head of the planning section was a white professional planner. The Board and various committees worked closely with the planners and their consultants, a few of which were companies composed of young blacks set up apparently for the sole purpose of being eligible to receive federal monies from the program.

After two years there were signs that the mayor and council's hands off policy had been interpreted as anything goes—and apparently anything went. An investigative team of reporters from the local newspaper were assigned to the case and received startling information from opinion leaders that had not been active in the Model Cities Program. Their research found that millions of dollars had been spent on planning, but not one dwelling unit, park, school, or street had been started. The staff of the program countered that the citizens wanted social services such as day care centers, legal services, job counseling, and job training rather than physical facilities which they labeled as white suburban frills. The reporters also found that most of the consultants were calling themselves planners but in fact appeared to be unqualified by virtue of experience or training. A few of the consultants were professional planners and were more or less coerced into hiring blacks to work on the program (few of which were qualified). Crime rates in the area were soaring, but there was no major examination of the problem. The situation in general was that millions had been spent but nothing had happened or even started to happen.

The federal government and the city were to conduct an evaluation of the program as a regular requirement of the federal funds. An intensive attitude survey was selected as the starting point by one of the competent professional planning counsultants. The survey was well structured and was an attempt to assess opinions on what was needed in the area by subgroups of the black population. The same survey was administered to the Board of Directors.

The results of the survey were rather astounding. The needs and wants of the people were aggregated to show that new houses, streets, parks, schools, and crime prevention programs were predominate. The results of the Board responses showed that day care centers, legal services, job counseling, and job training were predominate. The representatives of the people did not seem to be thinking what the people seemed to be thinking. When confronted with the results, the Board members took the defensive and refused to release the survey results claiming that the survey was irrelevant in a black community and was ingrained with racist value judgments. They also argued that only a 10 percent sample of four thousand people had been taken and thus there were thirty six thousand people not questioned who represented the true views. After sampling experts explained that statistical samples based on random selection were highly representative of the population, the Board became generally angry and terminated discussion on the evaluation and ordered the consultants paid and advised not to return to town.

The Model Cities Program went into a general decline at both the federal and local level. Funds were restricted and programs terminated with the rationale that, if the city really wanted the program, it could pay for it out of revenue-sharing funds. The mayor and council were in no mood to place revenue-sharing funds into the Model Cities Program and instead used the monies to roll back property taxes.

Several interesting lessons may be derived from this case. The obvious lesson is that town hall meetings are archaic and will only turn out those with special interests in the meeting, in this case a rather well organized group of welfare mothers. Any arguments that such mass conventions are representative of the people in the area are spurious. Since many of the special interest groups of the black and white community were denied access to the decision-making process through this means, including the mayor and councilmen, it was only a matter of time before the program would collapse. This implies that forthright attempts to identify opinion leaders by their special interest groups combined with the points of intervention strategy could have saved the Model Cities Program. It also implies that efforts and rhetoric concerning citizen participation by everyone in the community is not only infeasible but nonsensical. The utility of value analysis was demonstrated in this case, but the effectiveness of such analysis became dubious as constrained by the composition of the Board; hence the device enabled new insights and knowledge but was not sufficient on its own to force structural changes in the program. A final lesson is that representatives of the community already existed in the form of elected officials; and since the Model Cities Program systematically excluded their participation, there was no basis for implementation. Money was only available for planning.

NOTES

1. The most valuable study of political influence as it relates to planning is Edward C. Banfield, *Political Influence* (New York: Free Press, 1961).

2. A sound review of the expression and influence of public sentiments is available in Harlan Hahn, ed., *People and Politics in Urban Society* (Beverly Hills, Calif.: Sage, 1972).

3. An excellent survey of citizen participation in state and local government is available. See H. George Frederickson, ed., *Politics, Administration, and Citizen Participation* (New York: Intext, 1973).

4. James V. Cunningham, "Citizen Participation in Public Affairs," Public Administration Review, Vol. 32 (October, 1972), pp. 589-602.

5. Max Weber (trans. Don Martindale), *The City* (New York: Free Press, 1958).

6. This particular drivel was spewed forth by Phillip Arctander in "Dubious Dogmas of Urban Planning and Research," City, Vol. 6 (Winter, 1972), pp. 10-12, 54-56.

7. These four features were stated in an excellent article. See Richard S. Bolan, "Emerging Views of Planning," Journal of the American Institute of Planners, Vol. 33 (November, 1967), p. 234.

8. There have been many critics from diverse backgrounds making attacks on the Technocratic Form of citizen participation. Perhaps the most convincing and consistently accurate charges have been made by Herbert J. Gans. For example, see his: *People and Plans: Essays on Urban Problems and Solutions* (New York: Basic Books, 1968).

9. One of the few contemporary articles that defends the institutionalization of the Technocratic Form of planning is John T. Howard's "City Planning as a Social Movement, a Governmental Function, and a Technical Profession," in *Planning and the Urban Community,* Harvey S. Perloff ed., (Pittsburgh: University of Pittsburgh Press, 1961).

10. For a general analysis of the problems of agency self-interests, see Anthony Downs, *Inside Bureaucracy* (Boston: Little Brown, 1967).

11. These features are synthesized largely from the work of Paul Davidoff, especially his "The Planner as Advocate," in *Urban Government,* ed. Edward C. Banfield (New York: Free Press, 1969).

12. For a general discussion of these problems, see Cunningham, op. cit.

13. This point is made quite well by Lisa R. Peattie in her article "Reflections on Advocacy Planning," Journal of the American Institute of Planners Vol. 34 (March, 1968), pp. 81-86.

14. This general point is discussed more fully in Daniel P. Moynihan, "The Professionalization of Reform," in *The Great Society Reader: The Failure of Liberalism,* ed. Marvin E. Gettlemen and David Marmelstern (New York: Free Press, 1967).

15. This point was made strikingly clear by Martin Rein in his article "Social Planning: The Search for Legitimacy," Journal of the American Institute of Planners, Vol 35 (July, 1969), pp. 233-244.

16. Ibid., p. 234.

17. The genesis of this theory in its contemporary format can be linked to C. Wright Mills in his neoclassic *The Power Elite* (New York:

Oxford University Press, 1959). A transference of the theory (with a heavy dose of social change and liberal ideology) to planning is found in Alan S. Kravitz, "Mandarism: Planning as Handmaiden to Conservative Politics," in *Planning and Politics: Uneasy Partnership*, ed. Thad L. Beyle and George T. Lathrop (New York: Odyssey Press, 1970).

18. Mills, ibid., Chap. 2.

19. For a running commentary on how planners (although rather flippantly defined and broadly interpreted) tend to champion anti-capitalist and liberal groups rather than capitalistic and conservative groups, see Ayn Rand, ed., *The Ayn Rand Letter,* published fortnightly by the Ayn Rand Letter, Inc. For example, the November 20, 1972, letter (Vol. II, No. 4) has a concise commentary on how antipopulist many liberal politicians are despite their frequent use of such rhetoric as "Power to the People." Miss Rand argues that planners and liberal politicians will be overruled if the people really have the power to make their own decisions. These are somewhat extreme views, but then Miss Rand is the master of the subject.

20. While there have been many aspects of this research, a bench-mark effort is Elihu Katz, "The Two-Step Flow of Communication," Public Opinion Quarterly, Vol. 21 (Spring, 1957), pp. 61-78.

21. Cunningham, loc. cit., p. 599.

Chapter V

ON THE ROLES OF PLANNERS AND POLITICIANS

> *"The public must be served; and they that do it well, deserve public marks of honor and profit. To do so, men must have public minds, as well as salaries; or they will serve private ends at the public cost. Governments can never be well administered, but where those entrusted make conscience of well discharging their place."*
>
> –William Penn
> *Some Fruits of Solitude*

Not entirely tongue-in-cheek, one of the founding fathers of the planning profession offered his young proteges what he termed his "morning prayer":

"Oh Lord, give me the wisdom
to change the world today;
But let me use it
in a purely advisory way."

He said that he had incanted the prayer for many years, and his survival was its living testament.

The quintessential element of planning as a process, theory, technique, and profession in the American system of state and local government has been its advisory nature. Transformed into a role in the political process and life of state and local government, this has meant that the planner fulfills the role of advisor to the persons making the decisions, usually those representatives of the people who have been elected to public office. Periodically there have been suggestions from a surprising variety of sources that maybe the planner should do more than just advise. Perhaps the planners might be given limited delegations of political power which is equated with the power to make decisions. Such notions have been steadfastly repulsed by planners and politicians alike. Indeed those few that are in favor of limited delegations of decision-making by the politicians to the planners have taken to calling themselves radical planners in recent years.

Such an advisory role is like a smug and complacent icon to Nestor, a tribute to the imagery of seers and savants who avoid involvement and commitment. The elusive and puerile concept that the wisdom of the planner can stay alive above the fray of political life is nonsense. Some have argued that it is the absence of playing a role. Others have argued that it smacks of mandarinism in the sense that this means a condition of acting or behaving without being aware or conscious of the role really being played.[1] The historical obsession with planning is an advisory role to decision-makers as a satisfying and effective theory is illusory. There are alternate roles for planners and politicians which contain differing sets of decision-making authorities, and these alternates merit discussion and debate.

Executive Dominance Model

The basic roles established in the United States for political decision-making were invented during the Constitutional Convention in Philadelphia and eventually legitimized in organic law toward the close of the eighteenth century. Many of the founding fathers, like Thomas Paine, saw only two roles for government, that of making the laws and that of administering them.[2] The judicial role was seen as solely a part of the administration of laws. The federal government was designed and evolved into a tripartite role mechanism for political decision-making, however. The Congress would enact law; the Executive would administer law; and the Judiciary would be the final arbiter as to what was proper under the Constitution. In essence an all-powerful Judiciary was established in the United States which was quite unlike anything in Europe.

The basic three political roles set for the federal government are found to have filtered down to all of the states. In every state the three basic roles are legislative, executive, and judicial. Much more variety exists at the state level in that there are extraordinary patterns of permutations of power among the three roles. Like the federal government, however, there is increasing evidence that the chief executive is growing in political power. The long-term legacy of the Constitutional Convention appears to be that the demands of Alexander Hamilton for a superior chief executive may yet emerge in fact if not in form.

All powers to do planning, indeed even to exist as local political subdivisions, emanate from the state. Local government must originate, expand, and change with authorization and delegation from the state. Many states have given local governments ample powers to chart their own course through so-called home rule provisions of state constitutions or statutes, but the basic political powers nonetheless are state powers which can be delegated or taken back. The result of this situation has been a great variation in the roles set for decision-making in local government. The general role structure is still that of a legislative role and an executive role. Because of the higher courts at the state and federal level, however, the judicial role at the local level has dealt with details of law enforcement and has not played really a major role in political decision-making. The unique appeal system of the American judiciary almost guarantees that the higher courts at the state and federal level will play the more substantial role for political decision-making.

It can be stated, then, that the roles for politicians are essentially those of the making laws and establishing programs through the legislative process; administering and managing these laws and programs through the executive branch; and determining the legality and propriety of these laws and programs through the judicial branch. The executive and legislative roles are filled almost always by elected politicians. The judicial roles are filled by an assortment of elected and appointed politicians depending on the state. For present purposes, however, it is consistent to classify judges as politicians because they make decisions, effect checks and balances, and perform limited legislative functions.

These roles are compatible with the political process examined previously in that they all serve as checks and balances on one another, the basic role-playing strategy invented by the founding fathers. Suspicious of very powerful decision-makers in the European tradition, the founding fathers sought this system of stress which may require longer periods for decision-making but guarantees that there will be checks on power. The only way for these three roles to operate successfully in the political process is by compromise. Thus the roles for politicians in state and local government in the United States are inherently frought with stress, checks and balances, and compromise—but it works.

In light of these assertions, it is interesting to note that planning has existed in the United States as if the executive role were the only decision-making role for politicians; the role for planners is executive dominant. Most of the literature on planning, organizational theories, techniques and methods, and related aspects of role-playing for planners are oriented toward service as an assistant to decision-makers in the executive branch and even then primarily the chief executive. Perhaps this was not always so, but it clearly represents the pattern and trend for the role of planners since the end of the World War II. The basic role for planners in the tripartite role structure of state and local government thus emerges as an advisory aide to the executive role and observer to the legislative and judicial roles.

Many would argue that the heart of representative government at the state and local government is the legislative branch. Planners have had little interaction with the legislative branch in general, and much of the interaction has been adversary in nature. There is an unfortunate thread in planning thought which views the legislative branch as politics at its worst and politicians at their most reprehensible and vulgar. In reality the legislative branch is the most representative of the polity and political life in general.

The significant instances of interaction between the lawmakers and planners often occurs when the chief executive is weak because of organizational design or personality. In some local governments the mayor may be only a figurehead with the council and its committees performing the major roles for political decision-making. In such instances the planner is forced to deal with the legislative branch. Some states have very strong legislatures and inherently weak governors so that a similar situation occurs for state planning. In general, nonetheless, planners participate in the legislative branch decision-making only when that branch is performing a kind of executive role. The more typical legislative proceedings are devoid of planning input, however, at least from planners.

There is only a smattering of planning criticism and thought which argues for greater involvement with the legislative role of politicians. The most well known exponent of this argument goes so far as to suggest the restructuring of planning at the local level into a function of the local legislative body.[3] The proposal calls for breaking planning away from the executive branch and reorganizing it as a permanent staff to the legislative decision-makers for basic matters of laws, budgets, and programs. This arrangement, it is proferred, would bring planners into the real world of political decision-making and create a role that is broad and direct. While this is an intriguing posture, it has never been taken very seriously by most planners and no state or local government has adopted the suggestions precisely as stated. The argument is important, however, because it is one of the first to challenge the executive dominant role.

If closer linkages with the legislative branch have received scant attention by planners, closer linkages with the judicial branch have received virtually none. One of the rare proponents of greater roles for planners in the judicial branch bases the argument on the acceptance of the judiciary as an exercise in limited lawmaking. This means that the interpretation of laws is in essence the making of laws (which seems reasonable enough).[4] Such a role, while perhaps infringing upon some executive and legislative roles, requires the judiciary to take into account the future consequences of decisions. Since judicial decisions have effects on future conditions of state and local governments, and since these conditions are social, economic, physical, and political in nature, judges should have the same benefits of planning advice as legislators and executives. Unfortunately this and similar proposals go no further than that, but the notion is nonetheless interesting. It becomes even more interesting in light of trends in recent years to go to the courts after failing to obtain satisfaction in the executive and legislative branches, as has been done by some of the advocate planners. The entire argument for a greater role for planners in the judicial branch requires greater development however, as does the basic premise that the judiciary does and should perform quasi-legislative functions.

Despite the criticisms the executive dominant role for planners remains as the principal one for the profession. The theoretical basis for this role was established before World War II and found implementation shortly after that war.[5] The role was defined as that of the chief technical adviser to the chief executive on matters generally pertaining to the physical development of the state or local government. The role required the rendering of four basic services which were consistent with the basic seven step planning process described earlier. The first service to the chief executive was a definition of the problems and issues of the community in their technical context. The second service was to collect data about the problems and present the analysis to the chief executive. The third service was to present the alternative solutions to the problems to the chief executive and make recommendations on which was best. The fourth service was to provide continuing advice to the chief executive on the effectuation of the planning. In later years this executive dominant role was expanded slightly to include playing a role as agent of the chief executive to "educate" citizens on planning matters and "promoting planning among non-leaders who do not participate directly in the decision process."[6]

There are a number of inherent problems that plague the purely advisory role of planners within its executive dominant context. The most critical of these inherent problems is the placement of the planner in literally an adversary relationship with the legislators in the local councils or state legislatures. The legislative branch depicts the planners as the behind-the-scene and often invisible intellectuals who are putting bad ideas into the chief executive's head. It is not at all unusual for the legislators to mistake the planner as an evil influence upon the chief executive. Even worse, however, is the stereotype by the legislators of the planner as a sinister meddler into constituencies. Because the dislike between legislators and planners is often mutual, the adversary relationship not only continues but grows over time.

The advisory role of the planner in the executive dominant situation requires that the planner avoid any direct involvement with special interest groups or key individuals unless he is directed by the chief executive to cultivate such relationships. Since it is the style of most chief executives to reserve the direct communications for themselves, especially where there is publicity or high visibility, the planner usually must use surreptitious means to determine what are the objectives of the community, and he must be very careful in this undertaking. If the planner desires a more direct role in meeting with special interest groups, even on largely technical matters, the chief executive becomes wary. This wariness appears to vary directly with the relative power of the chief executive compared to the legislature or council. The planner cannot overtly complain or seek to change this low profile and stealthy posture because the executive dominant situation absolutely requires that the planner remain within the confidence of the chief executive. Only by doing always what is asked of him can the planner in such a role hope to be effective to any degree.

The executive dominant context may work in some situations where there is a capable chief executive who is sensitive and convinced that planning can be successful. There are indeed many examples where an able chief executive has used planning to the best degree within the constraints of the executive dominant situation.[7] There are of course other examples where an incompetent chief executive may or may not understand the value of planning and is helpless or unwilling to foster it. Therein lies the fatal flaw in the executive dominant model of planning: the model is based on the assumption that it is the chief executive who is always in charge of policy and decision-making. If the chief executive is not always in charge of policy and decision-making, the executive dominant model fails. Even if the chief executive is powerful but must share policy and decision-making responsibilities with the legislature and judiciary, as well as special interest groups and key individuals, the executive dominant model is bound to be of limited utility because it has built-in limitations for interaction. It can be rightly argued that the chief executive at the state and local level is at least as much in charge as anyone else, but those are not necessary and sufficient conditions for the executive dominant model to work.

The traditional home for planners is in the executive branch, and it is unlikely that the situation will change greatly regardless of the inherent problems. What is more likely to occur is a realignment of the planner's role to include more interaction with the other two branches of government, as well as with special interest groups and key individuals. The executive dominant model, in other words, has probably seen its day and is currently on the decline. This is a good sign because the executive dominant model for planning has been pushed to its logical extreme. The challenge for planning theorists, however, is to develop a philosophical basis whereby planners can serve the chief executive and the rest of the community. Such a role would obviously obfuscate the servant demeanor and install a managerial stance.

The Budget Cut Case

An interesting case study that concerns the executive dominant model can be regarded as either a success story or a lesson in failure depending upon how one views the long-term results. The setting is a large eastern state, and the case details have to do with the state and regional planning agency, primarily at the state level. Much of the impetus for the state and regional planning activities came directly from the governor who had been in office for a few years. The governor created a rather competent and fairly large staff which he lodged directly in his own offices. So confident was the governor in the state and regional planning agency that he lavished it with praise and continually sought to expand its functions and size. The agency became very powerful by virtue of this interpersonal relationship with the governor. The governor even used the potential for successful planning as a theme in his campaign for reelection, and won. Thus not only was a highly personalized executive dominant model of planning organized, but a governor with staying power was in office which offered unusual opportunities for success.

The initial programs of the agency were centralized and involved little if any participation by local governments or special interest groups and key individuals. A series of comprehensive planning documents were developed and released which sought to establish the long-range framework for state growth. A number of regional planning agencies were established by the governor to assist in the implementation of the comprehensive planning. As the agency grew in strength, largely as the result of direct support from the governor, it began to inject itself into many of the planning matters of the other state departments. After a few years of this astonishing climb, the state and regional planning agency had become the largest in the country with the best paid and most competent staff anywhere. It had acquired also an incredible number of enemies who were jealous of its privy relationship with the governor and fearful of its ambitions.

A difficult question which the agency faced at the azimuth of its growth was its role. Technically the state planning director was a member of the governor's cabinet and literally his right-hand man. Yet the governor insisted that the director and staff maintain a low profile and leave the highly visible aspects of state and regional planning to the governor and his political deputies. This created much confusion for both director and staff, yet they chose to suffer the ambiguity rather than propose changes in the role for planners. It seems fairly certain that the confusion in role and the low profile stance directed by the governor led the agency into nebulous areas and activities.

Supported largely by the opposing state agencies and some disgruntled local officials, the legislature proposed the abolition of the planning agency on the basis of its purportedly lavish and elitist staff, as well as its expensive budget. While the legislature argued that the agency had become fat and duplicative, several members were concerned equally with the apparent domination by the governor. The legislature sought to embarrass the governor by making it look as though the planning apparatus at the state level was a fiefdom with a single lord and many vassals. The governor was in an awkward positon and could not defend such a sophisticated agency in a year of budget cuts. His only alternative would have been to suggest cutting other areas that would have been politically disastrous. Reluctantly he watched the legislature dismantle the state and regional planning agency and was able to salvage only a modest staff. Most of the professional staff was released and the funds for planning reduced to a pittance. Throughout the entire battle, the planners maintained their stoic low profile and hoped for the personal strength of the governor to carry the day. Most of them became unemployed.

The state and regional planning agency in this case study had reached the top of the field in the eyes of most observers and is now in a slow and sporadic period of reconstruction. It is quite clear, however, that things will never be the same. The primary lesson that can be learned from this case study is that the executive dominant model has limitations and should not be pushed to extremes. While the executive branch is the proper home for the planning function, it must not be a prison for planners. The insistence on maintenance of a low profile by the chief executive meant that professionals could only feed ideas to politicos who fostered them with varying degrees of successes and failure. Such second-hand communications were bound to lead to misunderstanding and confusion for all parties involved. The governor may well have developed the elite and personalized planning he sought, but he should have realized its vulnerability.

As a good politician the governor should have realized the futility of isolating the planners from the legislators and special interests. Such isolation only serves to identify planners as convenient scapegoats when the political situation warrants. The planners themselves must share some of the blame, for they refused to argue forcefully for breaking out of this isolation and dealing forthrightly with the many groups concerned with state and regional planning. The planners contented themselves with being silent presences in a situation which called for openness and visibility. It is doubtful that any right thinking planner would ever opt for this kind of role again, if there is anything to be learned from experience and history.

The Growth Case

Another interesting case study offers some lessons to be learned with regard to the role that legislators can play in advancing planning programs when the chief executive fails to act. The setting is a southern state with a moderate governor who is commited to the furtherance of planning. The governor had sponsored a reorganization program which uplifted the state and regional planning function by incorporating the budget function within its responsibilities. The new planning and budget agency was placed in the executive offices of the governor and worked quite closely with his personal staff. As has been the common experience, the planners were asked to maintain a low profile and let the governor make the public appearances and high visibility presentations.

Neither the governor nor the planners had realized the immense amount of time required to prepare and present the executive budget document to the legislature. The underestimation resulted in the new agency spending almost all of its time on budget matters and somewhat letting slip the other aspects of planning. Both the governor and planners were not upset at this situation because they rationalized that the reorganization would require time to work the bugs out and this should be considered as an interim period. Among the various aspects of planning that were ignored during this time was the traditional concern of state government for growth and land use policy formation and control.

The growth and land use issue was of major concern to many special interest groups such as the conservationists and environmentalists. These special interests tend to include a number of prominent people who can be quite vocal when they want to be heard. They were concerned that the governor and his planners were spending little time on preparing growth policies and ways to control rampant and speculative growth. When they saw their suburban homes being surrounded by strip commercial developments and unbridled apartments catering to noisy and boisterous swinging singles, they began to get angry. Coalitions were formed and the areas of concern were broadened to include environmental protection, limitations on urban growth, and reform of zoning practices. Because the governor and planners were not responsive to these matters, the various special interest groups approached the legislators.

A group of moderate and informed legislators became sensitized to the growth and land use issues and realized that the governor had made a serious mistake by overlooking the need for immediate action on these matters. His steadfast and concentrated activities aimed at interrelating planning and the budget function were laudable but quite removed from the major concerns of the moment. A group of legislators approached the state planners and received inadequate responses to their questions on growth and land use and then decided to raise the issues on their own. A series of public hearings were held throughout the state and the degree of serious concern from special interest groups and key individuals was surprising to both legislators and planners. Few had realized the amount of anger and frustration that had arisen related to growth and land use in all parts of the state. Although it was probably not the intention of the involved legislators, the governor and his state planners received much more of the blame for the problems and inaction that they truly deserved.

The legislators were in a curious spot because they had raised an issue that was threatening to become explosive. The governor felt somewhat betrayed by his friends in the legislature and retreated in a defensive posture. The legislature had to respond, and a special drafting committee developed a major growth and land use bill which would require the state planners to designate the critical areas of the state for specialized land use controls, as well as develop a basic growth policy. Because there was much suspicion of the governor and state planners for their inactivity, however, most of the technical planning work was assigned to local and regional planning agencies. Needless to say the governor was not happy with this approach and most of the state planners felt that this was the wrong way to deal with the issues; they had been usurped nonetheless.

There are many things that could be said concerning this case study, but the more germane point is that the executive dominant model can fail even when the governor of a state is in a strong position with the legislature. Chief executives, like everyone else, make mistakes. When these mistakes are severe misjudgments of what are the major planning issues, there is a high likelihood that the inappropriate steps for rectification will be undertaken and planners will receive much of the blame for what has happened. There is a clear tendency for the executive dominant model of planning role playing to create a sheltered and removed aura for planners. They lose touch with special interest groups and other branches of government. This leads to a breakdown in communications and a high probability that the major planning issues will go unnoticed until a latent political situation occurs which rushes into the vacuum. When this happens it is certain to lead to embarrassment and loss of confidence in both the chief executive and planners.

Anti-Politics Factors

Many of the astute observers of planners and politics have argued that planners should assume a more activist stance in political matters. Some have gone so far as to say that the executive dominance model is a surrogate for real political activism and as such makes planning impossible in the American system of state and local government.[8] While there is no lack of clarion calls for political involvement by planners, there seems to be little in the way of specifics, however, and this tends to elicit a rather negative response set from planners.

Planners often respond to calls for political involvement by invoking the myriad of federal, state, and local laws that restrict the participation of public servants in partisan political activities. These anti-politics laws are repressive and misunderstood, which too often leads to their use as a catchall escape clause for political involvement. Indeed this can reach to the limits of the ridiculous, and there is even the story of the local planning director who once refused to make a recommendation on a controversial zoning change application because he said he was barred from such political activities by state law.

The impetus for anti-politics laws for civil servants was the assasination of President Garfield by a deranged office-seeker, which led to the Pendleton Act of 1883.

Most of the state and local anti-politics laws today are modifications or variations of the Hatch Act of 1939 which was passed by Congress to consolidate numerous rules made since 1883 and to help create a better merit system for federal civil servants. A 1940 amendment to the Hatch Act stated that state and local employees whose principal employment is related to programs receiving federal funds are included under the provisions of the Hatch Act. A 1949 amendment, however, excluded teaching and research employees for no detectable reason other than an ambitious lobbying effort by the involved special interest groups. State and local versions of the Hatch Act thus are redundant except where they impose even more restrictive provisions.

The Hatch Act is misunderstood by most civil servants in that the bar to political activities is defined in a manner that is more limited than generally assumed.[9] The Hatch Act prohibits campaigning, voter registration, circulation of nominating petitions, office-holding in political parties, fund raising, and candidacy in partisan elections. It does not prohibit expression of political opinions, advocacy of causes not limited to a single party, membership in political parties and groups, and full participation in "nonpartisan" political efforts and elections. Many public servants, unions, and professional groups argue that the Hatch Act is most likely unconstitutional because it violates the First Amendment to the Constitution. The U.S. Commission on Political Activity and Government Personnel appointed by the late President Lyndon Johnson concluded that the Hatch Act is unconstitutional and places an arbitrary and capricious set of restrictions on civil servants that has the effect of making them "second class citizens."[10] This commission recommended that Congress revoke the Hatch Act and replace it with a limited and definitive law protecting civil servants from forced contributions of time and money to candidates and parties but generally allowing them the same rights as the rest of the citizenry. A lack of organization by civil servant groups, as well as a prevailing sense of security in the existing Hatch Act by many of the old-time bureaucrats, has prevented any action on this proposal.

The Hatch Act came before the U.S. Supreme Court most recently in 1973. In fact, it has been before the Court regularly since 1947. The prehearing general feeling of optimism that this time the Court would assert its responsibility and strike down the Hatch Act as well as similar state and local laws evaporated with the upholding of the Act in a 6 to 3 decision.[11] Many observers believe that this decision was out of step with the times and anachronistic.[12] The Court held, in essence, that violations of freedom of speech were less substantive issues than the right of Congress to pass such laws. The government workers, and especially their unions, have become enraged at this decision and have been mounting increasing campaigns on Congress and the legislatures to revoke or reform anti-politics laws. These pressures should bear fruit in a few years.

The anti-politics laws perhaps were needed at the time of their enactment in a day when civil servants were a motley crew of political hangers-on and loyal party workers seeking the spoils of victory; however, those days are by and large gone now, and political life is vastly more complex. The tragedy of the anti-politics laws has been their misuse as a convenient cop-out by some planners to avoid intervention in controversial matters that usually were not partisan in nature anyway. Similarly these laws have been used as justification for not intervening in acts of the chief executive even when there were serious professional and technical questions involved. The anti-politics laws will be changed, and planners will not have a dubious rationale to justify their noninvolvement and value neutral role playing. This will be good for both the profession and state and local communities.

Alternate Political Roles: Individual

It is interesting to examine what alternative roles may exist for planners in light of the growing realization that the planner cannot remain as the isolated eunuch of the chief executive's office. Such an examination is needed since there is so little in the literature of the field that speculates on what kinds of political roles planners can play.[13] For purposes of clarity and simplicity, three roles, which reflect the principal variations inherent in the work of planners, can be discussed.

Apolitical-Neutral Role

The profession and practice of planning is institutionalized and technical to a sufficient extent to hypothesize that it will be possible for some planners to play an apolitical role that requires a value neutral stance on most matters. Such planners will most likely be involved with quantitative and technical aspects of planning, such as, modeling, information systems, simulation and gaming, and applied research. They may be involved also with routinized matters of planning such as administration of zoning, subdivision regulations, capital improvements, and traffic studies as long as non-controversial projects are involved. In many cases it is likely that the decision-making will be so mundane and regularized that state and local politicians will delegate some minor responsibilities to these apolitical-neutral role players, but there will be an exclusive right of review by politicians in order to deal with controversial matters and issues of substance that arise through some routine function.

There are several prominent features that can be attributed to such an apolitical-neutral role for planners. This role requires that even the appearance of political involvement or value judgment be avoided, and hence the planners will have to restrict their areas of concern in order to fulfill this condition. Similarly there should be no appearance of loyalties or confidences to any particular politician or political bloc. This would necessitate that planners fulfilling such a role be strongly bureaucratized and manifest any needs for strength and allegiance in terms of agency betterment. This bureaucratic posture will necessitate an organizational arrangement whereby the three branches of government have direct access to the planners and even the appearance of an executive dominant model is avoided. An important feature of this role is that the planner is to remain relatively anonymous and conscientiously avoid publicity. This will make it necessary for the planner to avoid contact with special interest groups and key individuals because such communications are likely to indicate that the subject being discussed is not a purely technical matter and thus beyond the province of this kind of planner.

The apolitical-neutral planner can enjoy the security, comfort, and life style of a bureaucrat and still fulfill certain limited functions of a technical nature which are necessary in state and local government. These will not be matters of great substance nor will they be matters that drastically change patterns and trends of growth and development. This kind of planner will not portend to be influencing policy and decision-making but will handle technical details of policy and decision-making. When this role places the planner in even an apparent confrontation with decision-makers, the apolitical-neutral planner should withdraw and offer no resistance to decision-makers. On the other hand the apolitical-neutral planner will have to develop highly structured and hopefully quantitative decision rules so that when he is dealing with his appropriate set of responsibilities he can always be consistent and predictable. Such a set of decision rules will avoid the situation where he is asked to compromise or make modifications since this is deemed to be a function of decision-makers.

The apolitical-neutral political role for planners is necessary for many of the activities of state and local government and is sufficiently definable and limited so that the appropriate set of responsibilities can be singled-out from other activities. The necessity for a high degree of technical method and standards is compatible with aspects of the Technocratic Form and even some variations of the Participatory Form.

While such a role may seem ingratiating and humiliating to some planners, it is nevertheless reflective of day-to-day activities that must be undertaken by persons with professional training and experience in planning. It is an appropriate role to be played by those planners who are technical by nature and dislike intensely the controversy and intrigue associated with the more substantive matters of planning that are meshed with politics.

Covert-Activist Role

There are certain conditions and contexts of a planning situation in which planners will be required to play generally the covert-activist role. This role is such that the planner will avoid the appearance of political involvement in general but will occassionally take positions and make recommendations that are obviously politically derived or based upon political compromises. The covert-activist role will be found in a bureaucratic situation where the planner is interested in both the betterment of the agency and the improvement of the community through intervention in key points of the political process of making decisions. Such a planner is likely to be in a position of responsibility that may be entitled director of the planning agency or a significant division or project within the planning agency. In most cases it is probable that the convert-activist planner will be cast in a typical executive dominant mold, but there will be aspects of legislative and judicial involvement. There may be some part of his work which is routinized and regularized in technical terms, but this kind of planner will not be as restricted in scope of activities as the apolitical-neutral planner.

The covert-activist usually will avoid the appearance of political involvement. He will express value judgments, loyalities, and confidences to certain decision-makers, as well as certain special interest groups and key individuals. In other words, the covert-activist planner will give the appearance of being a value neutral but will make his true beliefs clear to those special interests that are most likely to be useful in implementation of these values. Such a condition would require that the covert-activist planner make alliances with decision-makers and special interest groups but not get involved in the large-scale efforts at citizen participation except in a perfunctory way.

The covert-activist planner will sometimes foster special interest proposals but only when he is convinced that the common good of the community will not be diminished. He will always insist that the general public be served first, but may temper his stance with political realities on occasion. The rationale behind such a feature is that this kind of planner realizes the importance of agreement on substantive matters and does not hesitate to indulge in tradeoffs necessary for reaching agreements. He does not necessarily seek to build coalitions of special interest groups, but he is sensitive to the need for compromise and agreement prior to the general discussion of the issues. This condition also means that the covert-activist planner must gain entry into the offices of the decision-makers and opinion leaders. He can do so in part by frequently giving credit for planning successes to politicians and opinion leaders as representatives of political camps and special interests. Only rarely will the covert-activist role playing planner seek to take direct credit for successful planning.

The basic attitudinal feature that can be associated with the covert-activist planner is that the job of planning will take time and require steady and even-handed determination. He is a covert-activist because he does not necessarily seek to maximize the passions of the moment, preferring instead to intervene at key points of opportunity as part of a long-range strategy. He figures that he can at least outlast his political enemies by periodically submerging into technical matters. The covert-activist does not avoid technical matters but generally is aware of the limitations of the Technological Form. He is sensitive to the need for citizen participation, but he believes that it is right to impose his own value judgments when he is very sure that the common good of the community will be served.[14] Thus the covert-activist role is somewhat of a compromise condition that is compatible with aspects of the Technological, Participatory, and Activist Forms, but it cannot be neatly classified into any single approach to planning. The security and comfort of the covert-activist planner is less than the apolitical-neutral role, but more than the directly activist roles.

Overt-Activist Role

There is clearly a need in the planning profession for a group of overt-activist role players who can be identified as politically involved and action-oriented planners. These planners may or may not assume the traditional job titles and offices of planners, but at any rate they will function as the key advisers and leaders for planning matters. This may involve a partial delegation of some aspects of decision-making de facto although the formal decision-making process and roles will be unchanged. The overt-activist planner will seek to advocate successfully causes and values that are derived from the community (and deemed to be in the best interests of the community and its special interest components). The overt-activist planner will be visible to the polity and decision-makers and will be accountable for the results of the planning efforts. The overt-activist will sacrifice comfort and security and perhaps even life style in order to be more effective in formulating plans and programs and championing their effectuation.

The overt-activist planner will reflect often the policies and programs of the political leaders and principles with which he is identified. This will necessarily entail that he be involved in state and local political matters and have a keen sense of what is happening. This is not merely a role as a political errand boy, but instead this is a role of providing political leadership and direction. The overt-activist planner will hold certain values and principles which guide his planning efforts, and he will endeavor to make these clear to the community and its leaders. He will develop the needed confidences, loyalties, and mutualities that are deemed vital for successful planning. These features will require the overt-activist planner to be a mediator, builder of coalitions, conflict resolver, and manager of change–skills that are not easily acquired or held. The overt-activist planner will advocate causes and programs of special interest groups as long as they are consistent with the overall direction and intent of the planning efforts. The overt-activist planner will seek to learn what constitutes the public interest, but he will be realistic to the extent that plans cannot be based solely on that elusive criterion.

Is the overt-activist planner thus a power politician instead of a professional? The answer to this obvious question is no because the overt-activist planner does not run for office (at least not while functioning as a planner) and does not hold direct political power. The overt-activist planner may or may not enjoy political powers dependent upon the nature and scope of his confidences and loyalties with elected politicians especially the chief executive. To a limited degree he may play a role of power broker when certain competing special interests are vying for consideration in planning, but he will always base his power broker role upon values and principles that are inherent in the planning process he has defined for the community. In this capacity he may lose the anonymity enjoyed by the planners playing alternate roles, but he will usually seek credit for the causes with which he is identified rather than his own ego. In simpler terms, the overt-activist planner would know how to utilize the powers held by decision-makers without attempting to usurp those powers.

The overt-activist is still a professional planner and is not a professional politician. This requires that he be properly educated and trained and hold the appropriate professional credentials. He will be able to master and employ the various scientific methods and technical standards essential to the analytical aspects of planning work, but he will be aware of the constraints upon these tools. This condition will mean that the overt-activist planner uses discretion with regard to the sophistication, cost, and timing of the technical tools, always seeking to economically employ tools that offer the greatest potential improvement for decision-making. In most cases this often would mean that scientific methods and technical standards would be employed for the analysis of alternative policies, plans, and programs.

The overt-activist planner will differ greatly from the covert-activist and apolitical-neutral planners in that he will participate on occasion directly in political matters. These matters could well include campaign work and candidate support. Partisan political organizations and causes would be included within his scope of interests and involvements. In some cases he will stake his identity with certain persons and causes in a partisan context and will have to stake his future on such commitments. It may be necessary also that the overt-activist planner become involved in the nitty-gritty aspects of political activities which include fund-raising and management. Thus, in general, the overt-activist planner will differ from those playing alternate political roles in that he will truly be politicized in an open and straightforward manner which is devoid of hypocrisy and innuendo.

The direct politicization of the overt-activist planner indicates that the only form of planning with which he can be compatible is the Activist Form. While it is conceivable that certain variations of the Participatory Form could be channeled to make them compatible with the overt-activist role, it is quite unlikely that persons identified with the Technocratic Form of planning could play the overt-activist role.

Because the covert-activist and overt-activist roles entail a politicization of professional planners, it is well to discuss some problems associated with this feature. Prior discussion focused upon problems of nonpoliticization, but there are attendant problems with politicization as well.

Apolitical characteristics of professional planning are major problems that will have to be overcome. The aforementioned anti-politics laws will have to be changed and discarded in many cases, although there are obviously many loopholes for the politicized planner who earnestly seeks to be involved in political matters. There is also the anti-politics bias of planning theory and methods which must be changed. The covert and overt activist planners will somehow have to sift and winnow through their formal education in planning and utilize only those aspects that they deem most relevant. While that may not seem to be a major problem, it is because the anti-politic bias is so deeply ingrained in planning theory and thought that is often impossible to tell when it is present.

Another group of problems with politicization concerns the pressures and strains of political life. Political power carries a high price tag and many are unwilling to pay. The late President Harry S. Truman advised young people that political life was brutal and "if you can't take the heat then get the hell out of the kitchen." Such seemingly glib advice is startlingly correct. Many planners have become accustomed to an isolated and intellectual life style that precludes being the object of personal attack and criticism in many cases. Politicization of planning will eradicate such sanctuaries and refuges.

Related to the pressures and strains of political life are the inherent limitations on security and job longevity. Most political offices have time span limits and hence, if adaptations cannot be made, the planner must seek other employment. The problem is compounded by unforeseen election results and changing political alliances. It has been said that the most effective planners "keep their bags packed." Such a pessimistic piece of advice may be overly glib, but points out the reality of the dynamics of political life. Of course this dynamicism can have positive effects since the covert and overt planners can move on to higher levels of responsibility and trust although there are many who are not inclined to such movements (sometimes involuntary). The reality of this problem, however, is that there is no completely acceptable solution except in those places of single party rule and decision-makers with longevity; however, neither of these characteristics holds particular value for planning.

A different kind of problem that is related to growing political roles for planners is the work time and effort required for political activities. Few persons realize the amount of time, patience, and energy essential for political success. There are commitments to be made which are frightening to even the most stouthearted. This seems contrary to futuristic visions of a leisure society in which men graciously debate the issues of the day. Many planners may not desire to make such commitments for their time and energies, especially since the work of planners is quite heavy already. Other planners may find that they cannot continue to keep the pace necessitated by politicization. This problem also is insoluble in essence since the trends indicate that even more time and effort will be required for political life. A kind of Darwinistic effect may intervene here so that even though many planners may be willing to try the covert- and overt-activist roles, only a select group will have the wherewithal to make the grade.

A final problem worth mentioning is the dearth of educational experiences available for planners seeking the covert- and overt-activist roles. There is the basic question as to whether it will ever be possible to teach politics. These may be things which are best or only learned through experience. It may be possible to reorient educational programs so that the skills needed by such planners can be studied, however. This will be an obviously radical change for many planning schools which often have not shown evidence of being able to reform themselves. It is a quantum leap from studying land use plans to studying conflict resolution; there are great differences between zoning practices and learning coalition management; and there seems to be an eternity between design workshops and organizational behavior laboratories. The basic questions of whether planning schools and related educational programs can be directed toward the skills needed for politicized planners remain unanswered.

Three basic roles for individuals have been discussed above in the perspective of being the most identifiable and variegated. It is obvious that there will be cases when a given individual may exhibit features of each of these roles; thus, the roles are not entirely mutually exclusive as would appear on the surface. It also is possible that an individual planner may want to play varying roles at different times so that there is a dynamic aspect to these roles. In addition there are obvious mixes of the basic alternate roles that may be tailored for specific situations. Thus these qualifiers should be kept in mind when attempts are made to operationalize such roles for planners. In general, however, these three roles of apolitical-neutral, covert-activist, and overt-activist seem to be sufficient for examination of the roles possible for planners in the political process.

Alternate Political Roles: Organizational

The preceding discussion has been directed to planners as individuals within a political context. For various reasons there may be situations and times when planners must act as a group regardless of their position or rank in the political process. The group to which planners belong and are most likely to be responsible would be usually the American Institute of Planners and, in some instances, the American Society of Planning Officials. There may be related professional and scholarly groups, such as, the American Institute of Architects, American Society of Civil Engineers, American Society of Landscape Architects, American Society for Public Administration, National Association of Housing and Redevelopment Officials, and the like, within which planners may seek expression on various matters. There are also local groups and associations as well as political parties that are ad hoc or permanent to which planners may turn to for consideration. Whatever the particular group may be for a given issue, there are various aspects of planning and political interest wherein planners can act as a special interest group in the political process.

The basic advantage of working within the confines of an organized group for dealing with planning matters in their political context is the combination of anonymity for the individual and strength in numbers. This combination is conducive for obtaining insider's information and privileged information that would not be brought to the attention of the community in most cases or the decision-makers in some cases.[15] This information can be the deciding factor for many issues when used wisely by the particular group for which the planner is working. Some would glibly call this a "guerrilla" tactic, but in reality it is simply another outlet for the activist planner when he is constrained by his organizational identification. The special interest group offers such a planner both protection and a more formidable base of operations.

In some extreme cases it may be possible for the planner to act as a guerrilla by infilterating the planning agency in question and obtaining secret or other hidden information on how planning and decisions are undertaken. This can be within the structure of the objectives of the group for which the planner is acting but remain unknown to the planning agency under surveillance. In the most extreme cases the planner acting secretly for a special interest group may undertake acts of political sabotage meant to embarrass the enemies of the special interest group or expose these enemies to the community in an unfavorable political light. This guerrilla approach is found in only the rarest instances in contemporary practice, but the more critical consideration is that it does take place and shows that some planners (albeit extremely few) are acting radically out of value beliefs and judgments.

The more general utility of the group-oriented politicization of planners is for the fostering of reform and causes. These large-scale problems often require coalitions of special interest groups and are beyond the scope of the planner's activities. It thus becomes propitious for the planner to join in with larger groups of persons for seeking satisfaction and value attainments. It is becoming more and more apparent to many politicized planners that the most effective group for such activities is to be found within political parties. Working within political parties the planner can both provide information and technical know-how and participate in the establishment of partisan doctrine and beliefs. In this way the planner can work within the framework of the decision-making process that he would normally seek through advisory policy recommendations to decision-makers. If such a framework for decision-making finds its way into partisan campaign platforms, then there is the added benefit that perhaps there will be some binding effects upon elected party members.

The alternate political roles for planners that are available through special interest groups are not different in general from what they would be for any other member of the community. The essential difference is that the planner may have more to offer in some cases than the well-intended layman. This should not be viewed as a totally acceptable surrogate for politicization of the planner as an individual, however. While political roles through special interest groups may be effective for some planners, it will most likely be on an ad hoc, issue-by-issue basis rather than on a long-run pattern of success. In all but the most dogmatic and radical groups, this is likely to be inclusive of inconsistencies and changes that may be noisome to the ethics of many planners. Furthermore, the acting out of alternate political roles through special interest groups will not affect directly the more basic problems of the planning profession. This indicates that, while politicization for planners under the protection of special interest groups may be safer and more comfortable to some planners, if it is used to the exclusion of settling political problems for the profession, the planning profession will suffer. Alternate political roles for planners through special interest groups may win some battles but conceivably lose the war.

NOTES

1. Alan S. Kravitz, "Mandarinism: Planning as Handmaiden to Conservative Politics," in *Planning and Politics: Uneasy Partnership,* ed. Thad L. Beyle and George T. Lathop (New York: Odyssey Press, 1970). Kravitz transferred the concept of mandarinism to planners after examining the original utilization of the term to describe liberal intellectuals in general within the setting of the American foreign policy. See Noam Chomsky, *American Power and the New Mandarins* (New York: Free Press, 1969).

2. There are many good treatments of the principles debated during the Constitutional Convention. For example, J. H. Ferguson and D. E. McHenry, *The American System of Government* (New York: McGraw-Hill, 1959).

3. While well developed and stated convincingly this argument was actually only a part of a larger technical work on planning method. See T. J. Kent, *The Urban General Plan* (San Francisco: Chandler, 1964).

4. Robert K. Middleton, "The Planning Function in Government," a paper presented at the Planning Policy Conference, American Institute of Planners, Washington, D.C., February 21, 1973.

5. Generally accepted as the basis for the executive dominant role is Robert A. Walker's classic work *The Planning Function in Local Government* (Chicago: University of Chicago Press, 1941).

6. Robert T. Daland and John A. Parker, "Roles of the Planner in Urban Development," in *Urban Growth Dynamics: In a Regional Cluster of Cities,* ed. F. Stuart Chapin and Shirley F. Weiss (New York: Wiley, 1962), pp. 188-225. Quotation from p. 216.

7. One of the best known examples of a chief executive with an unusual affinity for planners has been Governor Nelson Rockefeller of New York who actually used the planning theme for a reelection campaign: "Rocky puts it all together." See Vincent J. Moore, "Poli-

tics, Planning, and Power in New York State: The Path From Theory to Reality," Journal of the American Institute of Planners, Vol. 37 (March, 1971), pp. 66-77. Shortly after the publication of Moore's work, the planning agency in New York fell onto hard times, as has been reported in Kathleen Agena, "Planning Metamorphosis in New York State," Planning, Vol. 37 (June, 1971), pp. 83-85, and in D. David Brandon, "Letter to the Editor," Planning, Vol. 37 (August, 1971), p. 114.

8. Thad L. Beyle and George T. Lathrop, "Planning and Politics: On Grounds of Incompatibility," in Beyle and Lathrop, op. cit., pp. 1-13.

9. Clyde Weaver, "Living with Hatch," Crucks, Vol. 2 (Fall, 1972).

10. U.S. Commission on Political Activity and Government Personnel, *Final Report* (Washington, D.C.: U.S. Government Printing Office, 1969).

11. United States Civil Service Commission, et. al., vs. National Association of Letter Carriers, AFL-CIO, et. al. 346 *F. Supp.* 578. The Court also refused to hear Broadrick, et. al. vs. Oklahoma, et. al. thus in effect upholding similar acts of states.

12. For an excellent analysis of the legal and moral issues involved, see Phillip L. Martin, "The Hatch Act in Court: Some Recent Developments," Public Administration Review, Vol. 33 (October, 1973), pp. 443-447.

13. This problem was discussed before a group of planners a few years ago and sparked much controversy and debate. See A. J. Catanese, "The Planner in Local Politics: An Activist Role," paper presented at the Annual Conference, American Institute of Planners, Minneapolis, Minn., October 19, 1970.

14. An interesting discussion of the correctness of planners imposing their beliefs on the community when the common good is clearly to be improved but the prescriptive medicine is bad to the taste is found in Herbert J. Gans, "The Public Interest and Community Participation," Journal of the American Institute of Planners, Vol. 39 (January, 1973), pp. 2-13.

15. A leading exponent of this approach has been Ralph Nader, "Consumerism, Advocacy, and Planners," paper presented at the Annual Conference, American Institute of Planners, Minneapolis, Minn., October 18, 1971.

Chapter VI

ON REACHING POSSIBLE DREAMS: CONCLUSIONS

> *"Ambition, avarice, personal animosity, party opposition, and many other motives, not more laudable than these, are apt to operate as well upon those who support, as upon those who oppose, the right side of a question. Were there not even these inducements to moderation, nothing could be more ill-judged than that intolerant spirit, which has, at all times, characterized politics."*
>
> –Alexander Hamilton
> *The Federalist, No. 1*

There is a compelling scene in Harold Robbins' steamy novel called *The Adventurers.* The super-hero, Dax Xenas, returns to his homeland, Corteguay, South America, after an absence of a few years. He had departed the country after his father led a revolution to overthrow a repressive dictatorship and install a democratic government. The democratic government had been withheld in favor of a strong centralized government headed by President Rojas who had been his father's friend but was now an enemy responsible for treachery.

In the scene, President Rojas needs Xenas' friendship because of his foreign capital sources and attempts to show him his successes.

"Our five-year plan calls for hospitals, schools, tractors, jobs, and food for all the people of Corteguay," boasts the President.

Xenas glares at the President and replies, "Yes, Corteguay will be a good place for the people—in five years."

This fictional account is somewhat similar to an expression attributed to Mao Tse-tung. Mao reportedly was reviewing the plans for China in the months following his victory. The planners were explaining to him that a country had to undergo certain periods of industrialization leading to the kind of post-industrial society enjoyed by such countries as Canada, the Soviet Union, and the United States. Such a phasing was possible and capable of being planned for economic, social, and political factors.

Deeply disturbed by this trend of thought, Mao replied, "Ten thousand years is too long to wait for the people of China."

The analogy to be derived from these stories is that planning is under attack from both the revolutionary and counterrevolutionary flanks. The revolutionary sees planning as an overly long and irrelevant exercise that serves as more of an excuse for inaction than for action. The counterrevolutionary sees planning as a means for slowing radical change yet is aware that planning raises the expectations of people to such a level that they will not settle for less than what has been promised (and sometimes demand even more). It would appear that planning is acceptable only to the middle-of-the-roader who can go either way and accept planning as either an effective or ineffective governmental activity depending upon the times and the issues at hand. The problem with such a philosophy is that the summation of the plays results in a zero sum game.

A far better philosophy is to reject the extremes of political philosophy (as well as the center) and consider planning as a governmental activity which serves as a catalyst for actions from various positions on the political spectrum. This is a view of planning as a means of transferring thought to reality. It would require basic changes in both planners and politicians but changes that are feasible. It would require planning to be feasible and practical rather than utopian and plausible. The following conclusions can be viewed as recommendations for two vital aspects of these changes. The first set of aspects concerns role and philosophical changes needed for the key actors in the political process. The second set of conclusions and recommendations is more attuned to the kinds of technical and procedural modifications needed for the planning process. Such a framework can be useful for determining the who, how, and when that is needed for planning within the politicized environment that is proposed.

Planners Should Be More Like Politicians

If planning is to be successful as measured by the quality of policies, plans, and programs developed and actions resulting from such efforts, then planners will have to adapt to more politicized roles because all of these activities occur within a political process. The political roles that appear to be practical and feasible are the aforementioned covert-activist and overt-activist roles for planners. There always will exist a demand for planners to fulfill the requirements of the apolitical-neutral role, but this role should be viewed as that of the technician and not confused with the roles necessary for meeting the greater potential of professional planning. The apolitical-neutral role is a retreat into technology while the covert and overt-activist roles are advances into the political process.

The planner should be more like a politician in that the real needs for such roles are certain skills that are attainable through professional training and experience. These skills are consistent with a systemic approach to planning in which the planner provides abilities to synthesize solutions to problems and employ skills which help to turn these solutions into real world actions.

The essence of the politicized role for the planner is such that he must reject the traditional self-justification as the savant of the community responsible for its total future and more properly view himself as the *manager of change.* Recent patterns quite clearly indicate that more and more responsibilities for planning and affecting the future are being placed in the hands of state and local governments instead of the national government (although the latter will monitor and fund such responsibilities). This means that more control of the destiny of the community is being placed within the hands of the community and its leaders. The planner can serve invaluably as the manager of the changes that will be generated by this decentralization and localization of governmental activities.

The manager of change conceptualization of the planner is a variation of the theory that holds government as the "conductor" of societal activities rather than the "doer."[1] The acceptance of this theory has been hampered by the inability to translate it into real world perspectives. By viewing planners as the managers of change who assist in the determination of the overall framework, which conforms generally to an integrated set of policies that will attain objectives set forth by the various special interest groups and key individuals of the community, such a theory can be made operational and satisfy many gaps that exist between planning and implementation. The overall framework that the planner will assist in developing includes long-, middle-, and short-range objectives, as well as day-to-day problem-solving. The planner is seen as the manager because he does not establish these objectives, or even the framework, by himself, but serves as the catalyst and expediter for the community in achieving such ambitions. This managerial function would require handling interrelations between the various special interest groups and key individuals, as well as setting the tone for community objectives. This role is not strictly managerial, however, because there will be innumerable occasions on which the planner will be asked to provide the leadership for the community in these matters. There will be a technical aspect of the manager of change concept because there will be many instances in which the community may be able to articulate objectives but will need the planner to formulate the specific plans, policies, and programs that are practical and feasible for reaching these objectives. The planner will act also as the key resource person when various supports and demands are presented in the political forum.

The shift from the doer of plans to the manager of planning within the political process would have the effect of removing planning from the backrooms and drafting rooms and bringing it into the streets and meeting places. It would mean that the planner would become, in the words of one writer, a "participant in the tremendous vitality of American pluralism."[2] As such, planning would lose much of its elitist imagery and become more identified as something that people do. The professional planners would lose their ivory-tower, intellectual auras and become more like champions and friends of the people.

Such ambitious matters for planners would necessitate that planners have something to offer that is not found among other professions. These skills may seem political but are actually managerial and technical skills (in addition to the synthesizing abilities of planners). The most important skill that the planner as the manager of change will need is that of conflict management.[3] As stated previously, the essence of the political system at the state and local level in America is stress and conflict as manifested in supports and demands for programs and policies. By managing the ramifications and conduct of this inherent conflict, the planner can offer what no other professional group can: practical and feasible results obtained through compromise and mutual adjustments. Compromise is the essence of conflict management and coalition building is the essence of compromise. The planners would help to hammer out the compromises, tradeoffs, incentives, and inducements that are needed for compromise. The planners would assist groups to understand the desirability of policies and courses of action held by other groups in order to achieve their own group interests. The planner as manager of change would be the conductor of conflict resolution related to the objectives of the community and would ensure that the results were just and fair to both the special interests groups and the community at large.

What specific skills would be needed by planners? As the manager of change utilizing the politicized role, the planner will need skills for (1) solving policy questions, (2) resolving group disputes and, (3) being effective through points of intervention in the political process. The solution of policy questions requires that the politicized planner have skills in mediation, negotiation, and bargaining. These are skills that can be studied formally and informally, as well as developed through experience and training. The resolution of disputes between special interest groups (and key individuals in many cases) necessitates that the politicized planner have the skills needed for solving policy questions as well as skills in group dynamics and interactions, social psychology, community organization, gaming and role playing, and basic management. These latter skills also appear to be among those that can be studied formally and informally, as well as developed through experience and training. The third basic set of skills is related to efficacy in points of intervention in the political process. In addition to the above skills and some basic common sense of politics, such skills are needed as systems approaches to planning, strategic policy analysis, quantitative methods, cost-benefit analysis, and opportunity costs analysis. This combination of political skills and quasi-scientific skills also can be studied formally and informally and acquired through experience and training. All three sets of skills must be kept within the overall perspective of the basic skill of the planner as a professional person who has abilities in synthesizing problems and their solutions. The planner can put it all together.

Why is this combination of skills within a politicized role for planners considered to be more political-like than professional? The basic justification for this consideration is that without it the context of the political process would be lost and the problems could be mistaken for technical and methodological rather than political and interpersonal. All conflict resolution for policies, plans, and programs in the public sector (as well as many in the private sector) that exists between special interest groups, individuals, and decision-makers will continue to take place within the political process discussed previously (or even with reform in a political process very much like the one described previously). The skills of the politicized planner are more political-like than professional because he will both work within this process and bring the crux of conflict resolution into the community. The planner and community would participate more adequately in the political process. The planner should consider the political perspectives of these skills also because he may be called upon to share some of the burdens of decision-making with the politicians as he grows in this new role and develops these skills.

Planners being more like politicians will require some structural changes within the planning profession. At the base of these structural changes will be a redefinition of planning and planners which is broader and less technical. This redefinition of planning and planners will have to deal with difficult questions pertaining to professional competence, education, training, and experience. The professional organizations for planners will have to redefine their purposes and objectives, as well as their codes of ethics, in order to be compatible with the proposed politicized roles for planners. It would seem appropriate also to recruit activists and representatives of special interests groups into the planning profession in order to show sincere commitments to changes and to develop greater credibility for the profession. In short, the politicization of planners will necessitate that more substance be given to the field of planning not through technical specialization only but through broadened definition and coverage of the political process at the state and local level.

Politicians Should Be More Like Planners

From the outset it has been maintained that improvements in planning at the state and local governmental levels could be attained only through changes in both planners and politicians. For planners in the politicized role to be effective, the decision-makers will have to demonstrate greater abilities in playing their proper roles in the political process. Some of the basis for development of these greater abilities by politicians can be derived from the inherent attributes of the concept of planning that will never be outmoded. Or to put it in another way, politicians should be more like planners in a number of ways.

The basic way that politicians should attempt to be more like planners is in their willingness to make commitments to planning as a means of reaching objectives set forth in the political process. There is remarkably little evidence that state and local politicians have ever made such a commitment although most will openly support the concept of planning and even admit to doing planning in their businesses and practices. Hence there exists the ironic feature that planning is accepted and acknowledged by state and local politicians but not followed. This can be rectified by the realistic adherence to the results of planning in the politicized vein by politicians even though this could mean some constraints on their bargaining leeways in many cases. The rationalization for such a self-imposed constraint is not just the better service possible to the whole community but the better service possible to the politician's constituency and favorite special interests since he will be able to bolster their input to the planning process.

Most politicians do not demonstrate an adequate confidence in planners and know remarkably little about how planning is or should be done. This can be interpreted as being caused by insufficient homework or other prepatory efforts by politicians. Many critics would argue that this simply shows that state and local politicos are stupid, but few critics ever get elected to public office. Politicians are not stupid, but may neglect the more complicated aspects of government, such as planning, in favor of more mundane daily problems and operations. It is in this aspect that a reordering of their interests to boost planning would be useful. It is useful also for politicians to become more personally acquainted and close to the professional planners in their employ if the requisite confidence is ever to be acquired.

Politicians should become more like planners by learning some new tricks of the trade with respect to the manner in which to use planners. Obviously the politicized planners will have skills, such as coalition-building, conflict resolution management, and negotiation, that are invaluable for politicians. Within the framework of the political process and the way that planners can relate to that process, it is conceivable that planners can and will work more closely with the individual politicians for solving many problems. This would have the dual payoff of helping the politician with his concerns and opening new points of intervention for the planner.

The politician can learn much from the way of thinking that planners have with regard to long-range implications of present-day actions and decisions. This is not to say that the politicians should adopt the archaic approach of putting everything off in favor of a thirty-year plan, but rather that the perspective of weighing actions and decisions against future conditions and probable consequences is crucial. Planners can assist politicians in this kind of thinking if the politicians are willing to receive such learning. This would mean that many politicians would have to reconsider their political philosophies regarding how they relate to their constituencies in the present and how they can best serve their constituencies now and in the future.

The biggest problem for all state and local elected leaders is how to relate and interact with their diverse, numerous, and complex constituencies. Clearly the newsletters, telegrams, polls, advertisements, and public hearings are not adequate for full interrelations. The older approaches of the neighborhood party clubhouse and ward leader, while superior to what exists today, are probably gone forever. The politician must develop new ways for determining the wishes and desires of his constituents, and little has been done in this area for decades. Perhaps the politician can learn from the planner how to create a viable and effective interrelation with his constituency by adapting some of the newer techniques of citizen participation in planning. The commitment of time and energy, as well as experimentation with new ways for increasing the visibility of elected leaders and their accessibility to the community and its special interest groups, is of paramount concern in this regard. Such a joint effort by planners and politicians could lead to satisfaction of constituents' demands for more expression of interests, as well as an improvement in the basis for planning and the basis for retention of political power.

There appears to be much merit in the extension of planning considerations into partisan positions and issues as well as individual campaign stances. The English system of partisan planning platforms and positions is superior to the nonexistant or window dressing approaches used in American politics. There should be Democratic and Republican planning proposals with substance and ideology. There should be planning issues raised on a partisan basis during election campaigns. Individuals should express their commitments and inclinations on planning matters as they run for office. In such a partisan mold it would be possible to make mutual partisan adjustments that would enable planning issues to be resolved. Avoidance of partisan positions leaves the resolution of planning problems to random or irrational factors that infiltrate the political process and are counterproductive to both parties and politicians.

A final recommendation for politicians to be more like planners has to do with the ability to retain an adherence to the supports and demands of special interest groups within the political process but improve capabilities for relating these supports and demands to larger aspects of the common good. This is not to say that the common good is easy to define; indeed, it is subtle and unpopular in many cases. The common good is something that politicians should be able to discern, however, through their intuitive and judgmental skills; it is for this reason that they are elected to represent others. The improved abilities to relate special interest group supports and demands to the common good could lead to benefits for the special interest groups that are in excess of the original intent through the myriad of spinoffs and indirect benefits that accrue in the system-like web of interrelations in the modern community. Such a posture requires the politician to be courageous and brave at many times since he will be following his intuitions on the long-run common good rather than the immediate satisficing proposals. Above all such a posture by the politician requires that he view his role in a historical sense. In this manner it could be argued that the politician can learn from the professional planner as he is emerging in thought and practice, and the result should be a closing of the gaps between planners and politicians.

Process and Method Improvements

Planners being more like politicians and politicians being more like planners formulate a basis for improved roles and philosophies in the political process. It is important equally to address the procedural and methodological aspects of planning because politicized planners are useless if they do not know what they are doing.

The process and method of planning suffers greatly in the present because it lacks substance and tends to be general and undefined. A roomful of planners have difficulty in defining what planning means. Methods are borrowed from other fields and disciplines. The old saw prevails: "Planning is what planners do." The situation can be compared to a problem from mathematics. In mathematics it is essential that a problem be defined, continuous, and limited if it is to be solved. A problem cannot be solved if it is ill-defined, discontinuous, and unconstrained.

Planning process and method are ill-defined because they differ greatly from place to place and practitioner to practitioner. They are discontinuous because there are ebbs and flows in funding, interest, and support for planning. Perhaps it would be possible to partially resolve the definition and continuity problems if the general planning process and method could be constrained. For purposes of argument, let it be assumed that the unconstrained planning process and method will always lead to frustrated efforts by planners. What is essential then is a set of anti-frustration constraints for planning process and method in order to make the politicized role and philosophy compatible with technical endeavors.[4] This set of anti-frustration constraints can be presented by listing the categorical problem of process and method and then the recommended constraint.

Taking Too Long

The actual elapsed time required for completing the procedural and methodological aspects of planning has been increasing consistently and has reached the point of being ridiculous. Inventories of data take so long to complete that the information is outdated when it is ready for use. Analytical methods of planning have grown so complex that years are required to utilize the models. The sum total of these exorbitant time sequences is to make planning such a lengthy process that it loses any capabilities for responsiveness and becomes bureaucratized.

The needed anti-frustration constraint is to reduce the time necessary for planning by establishing time limits for inventory and analysis and lessening the time of presentation. The tradeoff for the reduction in supposed sophistication of method would be the efficacy of results and responsiveness. Counter arguments that more and more time-consuming methods are needed because the problems are more complex are spurious because the data collected and analyzed are uniformly imperfect and should not be subjected to precise mathematical manipulations.[5] Clearly the time needed to do planning must be reduced.

Too Many Issues

Planners attempt to deal with too many issues, many of which are beyond their scope of competence. Many of the issues that planners spend much time on have not been recognized by politicians and special interest groups as being very important, and the planner often assumes the stance of raising issues that few care about. Some observers believe that planning for too many issues that are not considered critical by politicians and special interest groups within the community will result in nothing of value.[6]

The needed anti-frustration constraint is for the planner to deal primarily with those issues that have been deemed critical within the political process. If an issue is not considered critical, it does not merit immediate concentration. If an issue is invisible and appears to have no solution, it merits little planning attention. If planners are to be considered as good managers of change and responsive to both politicians and special interest groups, they must concern themselves with only those issues that are definable, concrete, and solvable. Governments, like people in general, have little tolerance for unclear, abstract, and seemingly insoluble issues that may concern no one but the few people studying them. Continued unconstrained planning attention to these latter issues would only serve to convince many critics that planners are intellectual esthetes little involved with the real world.

Too Many Promises

Planners have tended throughout history to promise too much and deliver too little. Even the critics of planning tend to conclude their arguments with the logic that if planning were only done better then most problems could be solved. Planning as the cure-all elixir for governmental and political ills is obviously a sham. Planning is but one aspect of the basic reforms needed for government. Planning as a process and method for dealing with all problems of government and politics is dubious because it is fairly evident that there may be some problems that cannot be planned for resolution. Indeed there is some common wisdom today which suggests that, instead of rationalizing every facet of society, it is better to leave some things to religion, fate, magic, hocus-pocus, and the great unknown.

The needed anti-frustration constraint for planning process and method is a self-constraint by planners in terms of promising less and delivering more. The planner must learn to say no, or at least maybe. The planner should exert self-control and avoid the natural tendency to respond positively to every request and demand placed before him. The planner should use this same self-control to avoid plunging into areas where he knows little of substance yet is called upon to provide answers. The result of this anti-frustration constraint should restore some credibility to planners' promises and place some asymptotic limits upon the levels of rising aspirations generated by planning and political promises that cannot be fulfilled.

Too Little Detail

Regardless of the type of plan, program, or recommendation offered by planners, there is a disturbing tendency to provide too little detail. Under the guise of such euphemisms as having plans that are flexible, dynamic, policy-oriented, and general, planners often manage to present plans that are devoid of detail and hence unusable. The rhetorical problem arises in that planners earnestly seek to instill notions of compromise and pragmatism in plans but extend the notions to an extreme by the avoidance of specificity and substance.

The needed anti-frustration constraint is for planners to insist on more detail and specificity in plans in order to make them useful for decision-making within the political context. This does not mean that plans need be inflexible, but rather that plans need to tackle the critical issues head-on. This does not mean that plans need be arbitrary, but rather that plans reflect the best judgment and experience of the planners. This does not mean that plans need be so picayune that every nuance of an issue be settled, but rather that the compromises available in the details be noted and presented. The net result of this insistence on greater detail in plans should be an improved basis for determination of tradeoffs, spinoffs, and by-products of recommendations so that the whole of the issue is better resolved.

Too Few Alternatives

To all but the most casual observer, it should be evident that planners have a tendency to violate the basic canon of planning process and method. This basic canon states that planners evaluate the alternative solutions to problems, perhaps not *all* but certainly the most feasible alternative solutions, and present these to the decision-makers. In real world practice the planners tend to present only those alternatives they like and in many cases no alternatives at all but instead one recommended set of propositions. This results in decision-makers and special interest groups concluding that there are several ways of doing something for which the planners always seem to have one other proposal.

If planning is to be at all effective within the schema of the political process that has been discussed, the needed anti-frustration constraint upon planning process and method will have to be a commitment to presenting the various alternatives that are feasible for solution. The planners must be able to give the decision-makers alternative solutions which are amenable to special interest group demands and supports. This may very well require that the planner bite the bullet at times and present an alternative solution that may be noxious to his sense of values and tastes but nonetheless feasible within the political environment in which he labors. More than a methodological and procedural question, it would appear that ample evaluation and presentation of alternative solutions is an ethical responsibility of the professional planner.

The imposition of these anti-frustration constraints upon planning process and method should be useful for dealing with the dilemmas expressed earlier by the Catanese Contention and its axioms. These dilemmas could be properly attacked by such constraints within the context of a politicized role and philosophy toward planning. In combination with recommended changes in role and philosophy for politicians, it would seem possible that the absolute impact of special interest groups could be somewhat reduced although admittedly never eliminated. Opening up the forum for the supports and demands that go into political decision-making will do much to help rectify the balance between special interest groups as opposed to the common good; but unless such reform is accompanied by basic changes in planning responsiveness, it would be meaningless.

Specific Improvements

There have been four more specific issues raised in previous discussion that have not been addressed directly in closing. These have to do with the rewards for planners, character issues, education, and professional organizations. These are the major specific issues that are somewhat detached from issues of role, philosophy, process, and method of planning.

Reward

The arguments for viewing planning as a profession based solely upon a tradition of selfless service to the people is losing its appeal and logic with time. It will be more and more difficult to justify a profession with so limited a horizon and a declining market. Americans are suspicious by nature of any group that insists upon public service as its only reward.

A radical new perspective would have to be added to the planning profession in order to deal with this issue in the long run. That perspective would take the form of viewing planning as a public service field for an interim period in a professional planner's career. Planners would go in and out of public service depending upon the results of elections and prevailing political moods and cycles. The long-run thrust of the planner's career should be in business, consulting, or education rather than public employment. The planner would expect his reward for good public service to accrue in his long-run career in the private sector (albeit often public-related).

This interim view of public service for planners would be consistent with the politicized role for all but the Technocrat Form of planning which has a market almost exclusively in public employment. Activist and Participatory Forms of planning would be compatible with the interim perspective of public service and should improve the efficacy of planners. This view of public service would not only help the rewards for planners but should assist in the larger common good through some reduction in bureaucracy. Finally the outlook of planners would be improved because a more mobile framework for employment could bring with it certain feelings of confidence whereby the courage can be mustered for making the tough decisions and fighting when a fight is necessary.

Character

While there do not appear to be major flaws in the character of planners, they are nonetheless different in major ways from politicians, which is probably good and bad at the same time. The more serious problem related to character is that planners have a bad public and political image which can only be changed by some character improvements.

The basic improvement would have to be that the character of planners is consistent with greater belief and commitment to the existing political and economic structure and order. Politicization of planners is not possible without such a commitment. This is not to say that planners should avoid trying to change things or advocating basic reforms. What is more specifically intended is to suggest a character stance such that all improvements and changes are possible and likely through working within the system, not destroying it or working from the outside. Quite clearly those who are dedicated toward utopian change and radical restructuring of political and economic structure and order should pursue their efforts in other professions or causes.

The image of planners with regard to their character needs much bolstering. The planners are viewed as idealists who are not action oriented or pragmatic. Stereotypes of image and direction abound from friends and enemies of planners. This character attack has little benefit and only results in less confidence in planners. Planners will have to make concerted efforts to improve their character images and may have to redefine some commitments in order to make these new images credible.

Education

There are so many reforms necessary in planning education that the topic should be treated as a separate entity. Nonetheless there are some long-run issues that merit present examination.

In the long run it may not be desirable to isolate planning education from other professional education. The continuation of the generalist with a speciality dogma that is already in its third decade of impotency cannot stand the test of time and will be rejected as an elitist conspiracy to reincarnate the Renaissance Man (who quite possibly never lived at that).

What is needed in the long run is the notion of the professional (or specialist) with knowledge of planning. It may be much more feasible to educate professionals about planning than generalists about professions. This would result in the planning education programs emerging as service centers on university campuses. They would offer advanced work and research as well. What would be gone in the long run is the notion of training persons who are supposedly capable of applying their skills to any problem area. In their place would be persons trained in management, politics, engineering, architecture, and other professions who have received additional training in the skills needed for planning which have been discussed previously.

The basic rationale for this proposal is but the failure of the existing model of planning education. This can be extended to include the concept that views of planning education occurring completely within the classroom are immeasurably naive. In the long run the best training for planners in both generalist and specialist senses is through experience outside of and beyond the classroom. By insisting upon some specialization for planners, properly identifying that specialization of planning can be better gained through experience and continuing education.

Professional Organization

It is quite likely that the American Institute of Planners will continue as the basic professional organization for planners at the state and local levels of government. No doubt related professional groups such as those for public administration, economics, sociology, and transportation will make inroads to the membership, especially among those planners practicing largely in these areas, but the Institute will continue to predominate nationally.

The Institute will have to make certain concessions in order to maintain its dominance of professional organization. It will have to broadly redefine the field of planning and ease up on its definitions for experience and education in an effort to attract more persons who are doing planning regardless of their educational and career origins. The Institute will have to decentralize its policy-making basis and reach more into state and local chapters for substantive policy-making. The Institute will have to spend a great deal of time on polishing its image and working more for the special interests of planners as a group which indeed would include some rough analogies to unionism. Such basic reforms would help to ensure the continued existence of the Institute in its primate position among planning professionals.

On Virtue

The reader was warned at the outset that many of these discourses would seem heretical and outside the virtue with which planners have been so long identified. These discourses have been concerned with *realpolitik,* the world as it is and will remain, and the analysis and proposals for planners have been so ordered. In conclusion it seems appropriate to restate the words of Machiavelli:

> ". . . the gulf between how one should live and how one does live is so wide that a man who neglects what is actually done learns the way to self-destruction. The fact is that a man who wants to act virtuously in every way necessarily comes to grief among so many who are not virtuous. Therefore if a prince wants to maintain his rule he must learn how not to be virtuous, and to make use of this or not according to the need."

NOTES

1. The foremost advocate, and perhaps the inventor, of this theory is Peter F. Drucker. See his *The Age of Discontinuity* (New York: Harper and Row, 1968).

2. This pithy statement of the role of politicized planners is found in Alan Altshuler, "Decision-making and the Trend Towards Pluralistic Planning," in *Urban Planning in Transition,* ed. Ernest Erber (New York: Grossman, 1970).

3. For an excellent essay on this point, see Dennis A. Rondinelli, "Urban Planning as Policy Analysis," Journal of the American Institute of Planners, Vol. 39 (January, 1973).

4. The concept of anti-frustration constraints upon planning process and method was first presented in two articles by this writer concerned with national planning in developing countries. See Anthony J. Catanese, "Planning in a State of Siege: The Colombia Experience," Land Economics, Vol. 40 (February 1973), and "Frustrations of National Planning: Theory and Reality in Colombia," Journal of the American Institute of Planners, Vol. 39 (March, 1973).

5. For an excellent technical argument against sophisticated mathematical models in planning, see William Alonso, "Predicting Best with Imperfect Data," Journal of the American Institute of Planners, Vol. 35 (July, 1968).

6. See Daniel P. Moynihan, "The City in Chassis," in *Toward a National Urban Policy,* ed. Daniel P. Moynihan (New York: Basic Books, 1970).

BIBLIOGRAPHY

BIBLIOGRAPHY

Ad Hoc Committee on Planning Education. *Final Report.* Washington: American Institute of Planners, 1973.

Agena, Kathleen. "Planning Metamorphosis in New York State," Planning, XXXVII (June, 1971).

Alonso, William. "Cities and City Planners," in H. W. Eldredge (ed.), *Taming Megalopolis.* New York: Praeger, 1967.

---. "Predicting Best with Imperfect Data," J. of the American Institute of Planners, XXXV (July, 1968).

Altshuler, Alan. *The City Planning Process: A Political Analysis.* Ithaca: Cornell University Press, 1965.

---. "Decision-making and the Trend Towards Pluralistic Planning," in Ernest Erber (ed.), *Urban Planning in Transition.* New York: Grossman, 1970.

American Society of Planning Officials. *Education and Career Information for Planning–1973.* Chicago: ASPO, 1973.

Anderson, Martin. *The Federal Bulldozer.* Cambridge: MIT Press, 1964.

Arctander, Philip. "Dubious Dogmas of Urban Planning and Research," City, VI (Winter, 1972).

Babcock, Richard F. *The Zoning Game: Municipal Practices and Politics.* Madison: University of Wisconsin Press, 1966.

Bair, Frederick H., *Planning Cities.* Chicago: American Society of Planning Officials, 1970.

Banfield, Edward C. *Political Influence.* New York: Free Press, 1961.

---. *The Unheavenly City.* Boston: Little, Brown, 1970.

Beyle, Thad L. and George T. Lathrop (eds.), *Planning and Politics: Uneasy Partnership.* New York: Odyssey Press, 1970.

Bolan, Richard S. "Emerging Views of Planning," J. of the American Institute of Planners, XXXIII (November, 1967).

Blecher, Earl M. *Advocacy Planning for Urban Development.* New York: Praeger, 1971.

Brandon, D. David. "Letter to the Editor," Planning, XXXVII (August, 1971).

Catanese, Anthony James. "Frustrations of National Planning: Theory and Reality in Colombia," J. of the American Institute of Planners, XXXIX (March, 1973).

———. "The Planner in Local Politics: An Activist Role," a paper presented at the Annual Conference, American Institute of Planners, Minneapolis, Minnesota, October 19, 1970.

———. "Planning in a State of Siege: The Colombia Experience," Land Economics, XXXX (February, 1973).

———. *Scientific Methods of Urban Analysis.* Urbana: University of Illinois Press, 1972.

Catanese, Anthony James and Alan W. Steiss. *Systemic Planning: Theory and Application.* Lexington: D. C. Heath & Co., 1970.

Chomsky, Noam. *American Power and the New Mandarins.* New York: Free Press, 1969.

Churchman, C. West. *Prediction and Optimal Decision: Philosophical Issues of a Science of Values.* Englewood Cliffs: Prentice-Hall, 1964.

Cunningham, James V. "Citizen Participation in Public Affairs," Public Administration Review, XXXII (October, 1972).

Dahl, Robert. *Who Governs?* New Haven: Yale University Press, 1961.

Daland, Robert T. and John A. Parker, "Roles of the Planner in Urban Development," in F. Stuart Chapin and Shirley F. Weiss (eds.). *Urban Growth Dynamics: In a Regional Cluster of Cities.* New York: Wiley, 1962.

Davidoff, Paul. "Advocacy and Pluralism in Planning," J. of the American Institute of Planners, XXXI (November, 1965).

———. "The Planner as Advocate," in Edward C. Banfield (ed.). *Urban Government.* New York: Free Press, 1969.

Downs, Anthony. *Inside Bureaucracy.* Boston: Little, Brown, 1967.

Drucker, Peter F. *The Age of Discontinuity.* New York: Harper and Row, 1968.

Easton, David A. *A Systems Analysis of Political Life.* New York: John Wiley, 1965.

Ferguson, J. H. and D. E. McHenry, *The American System of Government.* New York: McGraw-Hill, 1959.

Frederickson H. George (ed.). *Politics, Administration, and Citizen Participation.* New York: Intext, 1974.

Gans, Herbert J. *People and Plans: Esays on Urban Problems and Solutions.* New York: Basic Books, 1968.

———. "The Public Interest and Community Participation," J. of the American Institute of Planners, XXXIX (January, 1973).

Goodman, Robert. *After the Planners.* New York: Simon and Shuster, 1971.

Hahn, Harlan (ed.). *People and Politics in Urban Society.* Beverly Hills: Sage, 1972.

Harris, Britton. "Foreword," in Ira M. Robinson (ed.). *Decision-Making in Urban Planning: An Introduction to New Methologies.* Beverly Hills: Sage, 1972.

Howard, John T. "City Planning as a Social Movement, a Governmental Function, and a Technical Profession," in Harvey S. Perloff (ed.). *Planning and the Urban Community.* Pittsburgh: University of Pittsburgh Press, 1961.

Hunter, Floyd. *Community Power Structure.* Chapel Hill: University of North Carolina Press, 1953.

Judd, D. R. and R. E. Mendelson. *The Politics of Urban Planning: The East St. Louis Experience.* Urbana: University of Illinois Press, 1972.

Katz, Elihu. "The Two-Step Flow of Communication," Public Opinion Quarterly, XXI (Spring, 1957).

Kaufman, Herbert. "Emerging Conflicts in the Doctrine in Public Administration," American Political Science Review, L (December, 1956).

Kent, T. J. *The Urban General Plan.* San Francisco: Chandler, 1964.

Lee, Doublas B. Jr. "Requiem for Large-Scale Models," J. of the American Institute of Planners, XXXIX (May, 1973).

Lindbloom, Charles E. "The Science of Muddling Through," Public Administration Review, XIX (Spring, 1959).

Martin, Phillip L. "The Hatch Act in Court: Some Recent Developments," Public Administration Review, XXXIII (October, 1973).

Martin, Roscoe C. *Grass Roots: Rural Democracy in America.* New York: Harper and Row, 1957.

Meyerson, Martin and Edward C. Banfield. *Politics, Planning, and the Public Interest: The Case of Public Housing in Chicago.* New York: Free Press, 1955.

Middleton, Robert K. "The Planning Function in Government," a paper presented at the Planning Policy Conference, American Institute of Planners, Washington, D.C., February 21, 1973.

Mills, C. Wright. *The Power Elite.* New York: Oxford University Press, 1959.

Moore, Vincent J. "Politics, Planning, and Power in New York State: The Path From Theory to Reality," J. of the American Institute of Planners, XXXVII (March, 1971).

Moynihan, Daniel P. "The City in Chassis," in Daniel P. Moynihan (ed.). *Toward A National Urban Policy.* New York: Basic Books, 1970.

---. *Maximum Feasible Misunderstanding.* New York: Free Press, 1969.

---. "The Professionalization of Reform," in Marvin Gettlemen and David Marmelstein (eds.). *The Great Society Reader: The Failure of Liberalism.* New York: Free Press, 1967.

Nader, Ralph. "Consumerism, Advocacy, and Planners," a paper presented at the Annual Conference, American Institute of Planners, Minneapolis, Minnesota, October 18, 1970.

Peattie, Lisa R. "Reflections on Advocacy Planning," J. of the American Institute of Planners, XXXIV (March, 1968).

Perloff, Harvey. *Education for Planning.* Baltimore: John Hopkins University Press, 1957.

Planning: The ASPO Magazine. Vol. 38, No. 8 (September, 1972).

Rabinovitz, Francine F. *City Politics and Planning.* New York: Atherton, 1969.

Rand, Ayn. "Power to the People," *The Ayn Rand Letter,* II (November, 1972).

Ranney, David C. *Planning and Politics in the Metropolis.* Columbus: Charles Merrill, 1969.

Rein, Martin. "Social Planning: The Search for Legitimacy," J. of the American Institute of Planners, XXXV (July, 1969).

Rondinelli, Dennis A. "Urban Planning as Policy Analysis," J. of the American Institute of Planners, XXXIX (January, 1973).

Scott, Mel. *American City Planning Since 1890.* Berkeley: University of California, 1969.

Simon, Herbert A. *Models of Man: Social and Rational.* New York: Wiley, 1957.

Steiss, Alan W. "Planning and Decision-Making: Structural and Contextual Factors." Unpublished Ph.D. thesis, University of Wisconsin, 1971.

Walker, Robert A. *The Planning Function in Local Government.* Chicago: University of Chicago Press, 1941.

Weaver, Clyde. "Living with Hatch," Crucks, II (Fall, 1972).

Webber, Melvin M. "Urban Place and Nonplace Urban Realm" in Melvin M. Webber, et. al. *Explorations into Urban Structure.* Philadelphia: University of Pennsylvania Press, 1964.

Weber, Max. *The City.* Translated by Don Martindale. New York: Free Press, 1958.

Weiner, Norbert. *The Human Use of Human Beings: Cybernetics and Society.* Boston: Houghton-Mifflin, 1954.

Woodbury, Coleman (ed.). *The Future of Cities and Urban Redevelopment.* Chicago: University of Chicago Press, 1953.

INDEX

INDEX

ABOUT THE AUTHOR

ANTHONY JAMES CATANESE is a teacher, consultant, and writer on urban planning and management. He has three degrees in the field, including a Ph.D. from the University of Wisconsin. Currently at the University of Miami, he has taught previously at the Georgia Institute of Technology, University of Wisconsin, Virginia Polytechnic Institute and State University, Clark College, and Pontificia Universidad Javeriana, Bogotá, Colombia. He is a national officer in the American Institute of Planners. His other books include *Systemic Planning: Theory and Application, New Perspectives on Urban Transportation Research,* and *Scientific Methods of Urban Analysis;* and he has contributed extensively to professional and scholarly journals and magazines.

Dr. Catanese has participated directly in state and local politics. He has been a successful candidate for several offices in local government, and has been a key campaigner for several national, state, and local candidates.

NOTES